MATHEMATICS

Complied and Edited by:
Yule Bricks

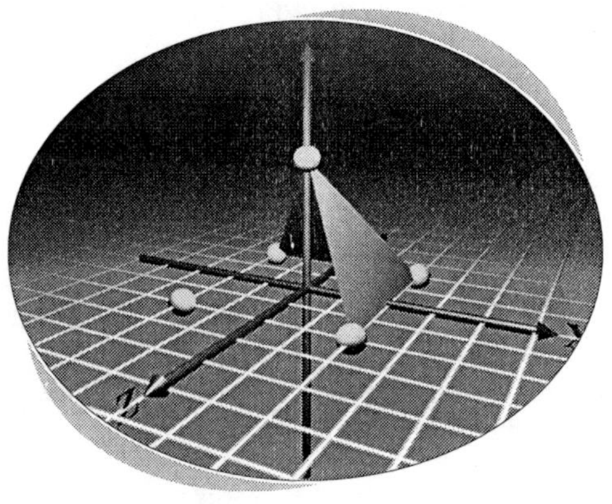

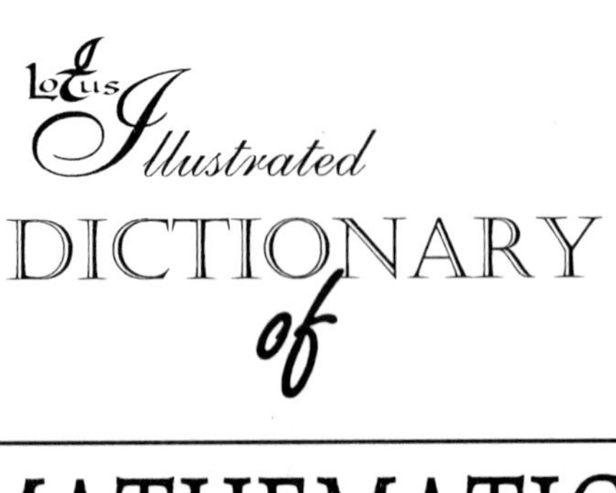

MATHEMATICS

ISBN 81 89093 47 9

Published by:

Lotus Press : Publishers & Distributors
Unit No. 220, 2nd Floor, 4735/22,
Prakash Deep Building, Ansari Road, Darya Ganj,
New Delhi- 110002, Ph.: 41325510, 09811838000
• E-mail : lotuspress1984@gmail.com www.lotuspress.co.in

Printed & Published by : **Lotus Press Publisher & Distributors,** New Delhi-02

PREFACE

Every culture on earth has developed some mathematics. In some cases, this mathematics has spread from one culture to another. Now there is one predominant international mathematics and this mathematics has quite a history. It has roots in ancient Egypt and Babylonia, then grew rapidly in ancient Greece. Mathematics written in ancient Greek was translated into Arabic. About the same time some mathematics of India was translated into Arabic. Later some of this mathematics was translated into Latin and became the mathematics of Western Europe. Over a period of several hundred years, it became the mathematics of the world.

There are other places in the world that developed significant mathematics, such as China, Southern India and Japan and they are interesting to study, but the mathematics of the other regions have not had much influence on current international mathematics. There is, of course, much mathematics being done these and other regions, but it is not the traditional math of the regions, but international mathematics.

The science of patterns and order and the study of measurement, properties and the relationships of quantities, using numbers and symbols. There are different branches of mathematics like algebra, trigonometry, differential calculus, integral calculus, differential equations, analytic geometry, plane geometry and solid geometry statistics.

As our knowledge increases inexorably Mathematics is taking an increasingly important part in modern society and as computers and the use of statistical techniques now pervade most areas of life, familiarity with mathematical ideas.

Newton predicted the motion of the planets; be immersed in *'the Calculus'* which is man's most powerful mathematical tool and underlies almost every aspect of human endeavour. For Mathematics is behind virtually every aspect of life: if you ask 'why' persistently enough the final answer will be a mathematical one. So the aptitudes needed for mathematics are essentially a deep curiosity, an interest and sufficient ability.

This book contains all the terms used in solving any calculation. It includes description of basic geometrical figures, theorem, derived laws and mathematical symbols, making it a perfect guide for the mathematics enthusiasts world over.

a posteriori

reasoning which proceeds from effects to causes; which deduces general rules by looking at patterns is called a posteriori reasoning. A statement whose truth can only be found out by experience is also called a posteriori.

abacus

a Japanese counting device and calculator.An ancient calculating device made of beads and wires mounted on a frame. Often used to teach place value.

abelian group

a group in which the binary operation is commutative, that is, ab=ba for all elements a abd b in the group.

abscissa

the x-coordinate of a point in a 2-dimensional coordinate system.

absolute value

1. the positive value for a real number, disregarding the sign. Written $|x|$. For example, $|3|=3$, $|-4|=4$, and $|0|=0$.

2. the distance away from the origin. This means the positive value of a number. For example, the absolute value of -6 is 6.

abstract number

a number with no associated units.

abundant number

a positive integer that is smaller than the sum of its proper divisors.

acceleration

the rate of change, the derivative of velocity. If position is represented by s(t), then velocity is s'(t) and the acceleration is s''(t). The change in velocity divided by the change in time.

■ acute angle
an angle which measures less than 90 degrees.

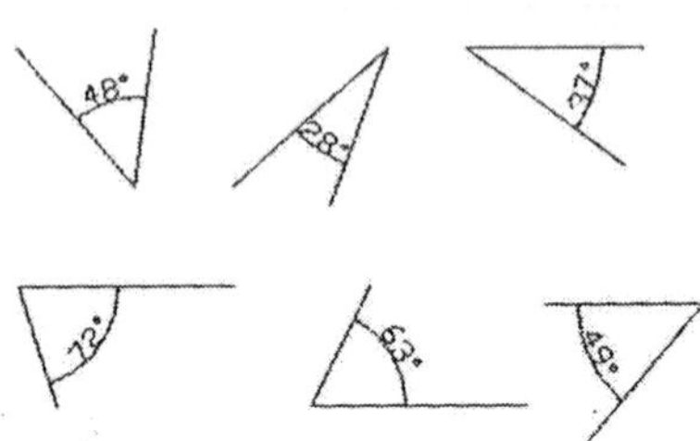

■ addend
the numbers being added in an addition question.

■ addition
the process of finding the sum of two numbers, which are called addend and the augend (sometimes both are called the addend).

■ additive identity
the number zero is called the additive identity because when you add it to a number, N, the result you get is the same number, N ... 0 + 5 = 5.

■ additive inverse
the additive inverse of a number, N, is the number that when you add it to N, the result that you get is zero. The additive inverse of 7 is -7 ... 7 + -7 = 0.

■ additive inverse
two numbers that equal 0 when added together. The numbers 14 and -14 are additive inverses of each other.

■ adjacent angles
two angles which share a common side and the same vertex.

■ affine transformation
see transformation.

■ algebra
a branch of mathematics in which variables are substituted for unknown values to solve a particular problem.

■ algebraic equation
an equation of the form f(x)=0 where f is a polynomial.

■ algebraic geometry
traditionally, the geometry of solutions in the complex numbers to polynomial equations. Modern algebraic geometry is also concerned with algebraic varieties, which are a generalisation of such solution sets, as well as

solutions in fields other than complex numbers, for example finite fields.

algebraic methods
the use of symbols to represent quantities and signs to represent their relationships.

algebraic number
a number that is the root of an algebraic polynomial. For example, sqrt(2) is an algebraic number because it is a solution of the equation $x^2=2$.

algebraic sentence
a general term for equations and inequalities.

algebraic topology
the branch of topology concerned with homology and other algebraic models of topological spaces.

algebraic variety
a space which is locally the solution locus to a set of polynomial equations. Algebraic varieties are for algebraic geometry{algebraic geometers} what topological spaces are for topology{topologists}. Indeed, many algebraic varieties are (complex) manifolds. However, algebraic varieties may also have complicated singular sets and may be parametrised with rings other than the complex numbers. (For the technical reason that the real numbers are not algebraically closed, one does not consider algebraic varieties over the real numbers in the straightforward sense.).

algorithms
a mechanical procedure for performing a given calculation or solving a problem through step-by-step procedures such as those used in long division.

almost complex manifold
a manifold with the property that each tangent space has the structure of a complex vector space, but the complex structures are not necessarily compatible with true complex coordinates as they are for a complex manifold.

alphametic

a cryptarithm in which the letters, which represent distinct digits, form related words or meaningful phrases.

alternate hypothesis

an alternative hypothesis is one that specifies that the null hypothesis is not true. The alternative hypothesis is false when the null hypothesis is true, and true when the null hypothesis is false. The symbol H1 is used for the alternate hypothesis.

alternating-sign matrix

a matrix of 0's, 1's, and -1's such that, if the zeroes are deleted from any row or column, the remaining entries alternate in sign and begin and end with 1.

alternative hypothesis

in hypothesis testing, a null hypothesis (typically that there is no effect) is compared with an alternative hypothesis (typically that there is an effect, or that there is an effect of a particular sign). For example, in evaluating whether a new cancer remedy works, the null hypothesis typically would be that the remedy does not work, while the alternative hypothesis would be that the remedy does work. When the data are sufficiently improbable under the assumption that the null hypothesis is true, the null hypothesis is rejected in favor of the alternative hypothesis. (This does not imply that the data are probable under the assumption that the alternative hypothesis is true, nor that the null hypothesis is false, nor that the alternative hypothesis is true.

altitude

1.the altitude of a triangle is the line segment from one vertex that is perpendicular to the opposite side.

2.the distance from the base of a figure to the highest

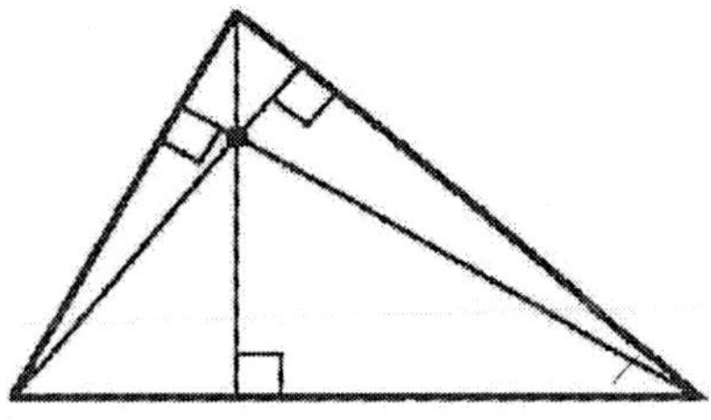

point in the figure, especially in a triangle.

■ **amicable numbers**

two numbers are said to be amicable if each is equal to the sum of the proper divisors of the other.

■ **amplitude**

half the distance from the highest point to the lowest point. The amplitude of a normal sine curve is 1, since the distance between top and bottom is 2.

■ **analysis**

one of the three traditional branches of mathematics, along with algebra and geometry. It is the branch concerned with estimates, inequalities, differential and integral calculus, and properties of the real numbers. It includes such areas as real analysis, functional analysis, operator theory, measure theory, differential equations, and special functions.

■ **analytic**

in analysis, a function or a structure described by functions is analytic its Taylor series converges to it. Since the function must first have a Taylor series, it is in particular smooth if it is analytic.

■ **angle**

1. the figure formed by two line segments or rays that extend from a given point.

2. the measure of the distance between two values that share a common point (measured in degrees).

■ **angle measure**

the measure of the space between two lines that meet in a point. Angles are measured in degrees or radians.

■ **annulus**

the region enclosed by two concentric circles.

■ **ante**

the up-front cost of a bet the money you must pay to play the game. From Latin for "before."

■ **antecedent**

the hypothesis of conditional statement. The "if" part of an "if-then" statement.

antiderivative
the antiderivative of a function, f(x), is a function, F(x), whose derivative is f(x). Also called the indefinite integral.

apothem
the perpendicular distance from the center to a side of a regular polygon.

applet
an applet is a small program that is automatically downloaded from a website to your computer when you visit a particular web page; it allows a page to be interactive—to respond to your input. The applet runs on your computer, not the computer that hosted the web page. These materials contain many applets to illustrate statistical concepts and to help you to analyse data. Many of them are accessible directly from the tools page.

applied mathematics
mathematics for the sake of its use to science or society.

arc length
s = integral (sqrt (1 + $(dy/dx)^2$)) dx.

Archimedes(born 287 B.C.)
Greek geometer and applied mathematician who lived at Syracuse, a Greek settlement on the coast of Sicily. He is famous both for his profound mathematical work and for his ingenious mechanical inventions, which at the end of his life he used to defend his city against the Romans during the Second Punic War. When he discovered the hydrostatic principle – that a floating body of a given weight displaces a volume of liquid having an equal weight – he is reputed to have jumped from his bath and run naked through the streets shouting "Eureka!" (I have found it!) Archimedes is credited with many brilliant mathematical discoveries, especially in properties of geometrical figures, areas and volumes of conics, and principles of trigonometry.

(He is thought to have known Heron's formula – several centuries before Heron.) He found areas of plane figures by the method of exhaustion, prefiguring modern integral calculus. A Roman soldier killed him during the sack of Syracuse.

■ **area**

the amount of space taken up by a two-dimensional figure.

■ **argument**

the independent variable in a function.

■ **Aristotle(born 384 B.C.)**

greek philosopher and scientist who developed syllogistic logic as a formal scholastic discipline. He defined the syllogism as a "discourse in which certain things having been stated, something else follows of necessity from their being so." Aristotle's philosophy of the infinite, in which he held that only the potential infinite, and not any collection which was actually infinite, could be entertained by the reason, held great authority for 2,000 years, until Georg Cantor introduced his theory of transfinite sets in the 19th century. Aristotle wrote widely on every field of learning, interpreted and disagreed with the writings of his teacher Plato, and served as tutor to Alexander the Great.

■ **arithmetic**

the branch of mathematics dealing with the rules for operations on numbers.

■ **arithmetic mean**

1. the arithmetic mean of n numbers is the sum of the numbers divided by n.
2. the average of a set of numbers. The arithmetic mean of 8 and 4 is (8+4)/2, or 6.

■ **arithmetic mean**

the sum of a set of numbers divided by the number of numbers. Also called the average.

■ **arithmetic sequence**

a set of elements ordered in some specific way (e.g., {2, 5, 8, 11, 14, ...}) in which there is a common difference between successive terms. As in the example above, where 3 is the common difference. Each successive term of the arithmetic sequence is found by adding the common difference to the preceding term.

arithmetic series
the sum of an arithmetic sequence. Ex Given the arithmetic sequence 3, 5, 7, 9, 11 ... the arithmetic series would be 3 + 5 + 7 + 9 + 11.

aronszajn tree
for α a regular cardinal, an α-Aronszajn tree is a tree T with |T| = α such that every chain in T is of cardinality less than α.

array
an orderly arrangement of objects in columns and rows.

Ascoli-Arzelà theorem
if {X} is a compact Hausdorff space and F is an equicontinuous, pointwise bounded subset of the space C(X) of continuous functions on X, then F is totally bounded in the uniform metric and the closure of F in C(X) is compact. Also, if X is a σ-compact, locally compact Hausdorff space and if (f_n) is an equicontinuous, point wise bounded sequence in C(X), then there exists a subsequence and an f in C(X) such that the subsequence converges to f uniformly on compact sets.

association
two variables are associated if some of the variability of one can be accounted for by the other. In a scatter plot of the two variables, if the scatter in the values of the variable plotted on the vertical axis is smaller in narrow ranges of the variable plotted on the horizontal axis (i.e., in vertical "slices") than it is overall, the two variables are associated. The correlation coefficient is a measure of linear association, which is a special case of association in which large values of one variable tend to occur with large values of the

other, and small values of one tend to occur with small values of the other (positive association), or in which large values of one tend to occur with small values of the other, and vice versa (negative association).

■ **associative**

when three or more numbers are added or multiplied, the operations can be performed in any order.

e.g., (3 + 4) + 6 = 3 + (4 + 6).

■ **associative property**

a binary operation * is defined associative if, for a*(b*c) = (a*b)*c. For example, the operations addition and multiplication of natural numbers are associative, but subtraction and division are not.

■ **associative property of addition**

(a + b) + c = a + (b + c).

■ **associative property of multiplication**

(a * b) * c = a * (b * c).

■ **asymptote**

a straight line that is a close approximation to a curve as the curve is drawn; in other words, a line that a curve or function gets close to but never crosses. The two most common types of asymptotes are horizontal and vertical.

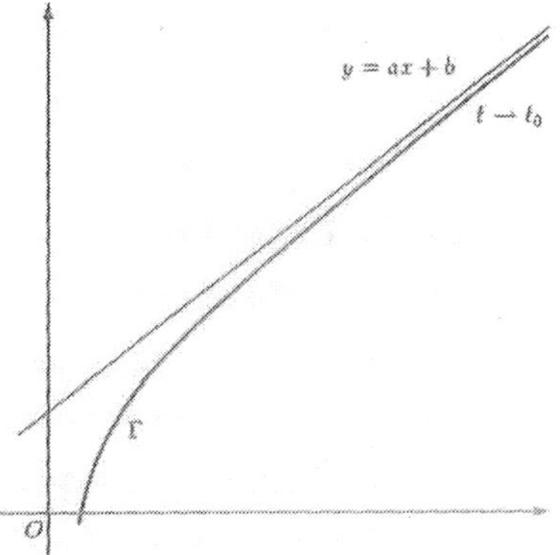

■ **asymptotics**

a general term in mathematics referring to the properties of an object as key parameters such as the dimension, the non-linearities, or the length scale, become very large or very small.

■ **Attouch-Wets topology**

a certain topology on functions or compact sets on an infinite-dimensional vector space which is useful in analysis and optimisation problems. A sequence converges in this topology if it converges (uniformly, or in the Hausdorff metric) in each finite ball centreed at the origin.

attribute

quality or characteristic.
e.g., the attribute of mass is weight; an attribute of a block is its thickness.

Aubel's theorem

given a quadrilateral, if we draw a square on each side the two lines joining to the centres of squares on opposite sides are perpendicular, and of equal length.

August Möbius (1790-1868)

a German mathematician, now remembered mainly for the Möbius strip which is named after him. He worked in astronomy as well as trigonometrical equations, geometry and projective geometry. The Möbius transformation is named after him. He was interested in topology, and worked on questions to do with map-colouring, although the Möbius strip was in fact discovered by a mathematician named Listing.

automorphic number or automorph

a number n whose square ends in n. For instance 5 is automorphic, because 52 = 25, which ends in 5.

automorphism group

the automorphisms of any object or structure can be composed to produce new automorphisms of that object. Since functional composition is associative, and since automorphisms are by their nature invertible, the automorphisms of an object or structure form a group (with the trivial automorphism serving as the identity element). This group is called the automorphism group of that object, usually denoted by Aut(X) for some object X. Automorphism groups are very useful, as they yield information about the symmetries of objects at hand.

average

a single number that gives a measure of central tendency of the numbers in a set. Usually, this is the mean. (It could also be the median or the mode.).

avoirdupois

a British system of units of mass, based on the pound:
7000 grains = 1 pound.
16 drams = 1 ounce.
16 ounces = 1 pound.
14 pounds = 1 stone.
2 stones = 1 quarter.
4 quarters = 1 hundredweight.
20 hundredweight = 1 ton.

axiom

a statement that is assumed to be true without proof. Postulate.

axiomatic systems

systems that include self-evident truths; truths without proof and from which further statements, or theorems, can be derived.

axioms of probability

there are three axioms of probability (1) Chances are always at least zero. (2) The chance that something happens is 100%. (3) If two events cannot both occur at the same time (if they are disjoint or mutually exclusive), the chance that either one occurs is the sum of the chances that each occurs. For example, consider an experiment that consists of tossing a coin once. The first axiom says that the chance that the coin lands heads, for instance, must be at least zero. The second axiom says that the chance that the coin either lands heads or lands tails or lands on its edge or doesn't land at all is 100%. The third axiom says that the chance that the coin either lands heads or lands tails is the sum of the chance that the coin lands heads and the chance that the coin lands tails, because both cannot occur in the same coin toss. All other mathematical facts about probability can be derived from these three axioms. For example, it is true that the chance that an event does not occur is (100% - the chance that the event occurs). This is a consequence of the second and third axioms.

■ axis of a pyramid

the axis of a pyramid is the line joining the top of the pyramid to the centre of its base.

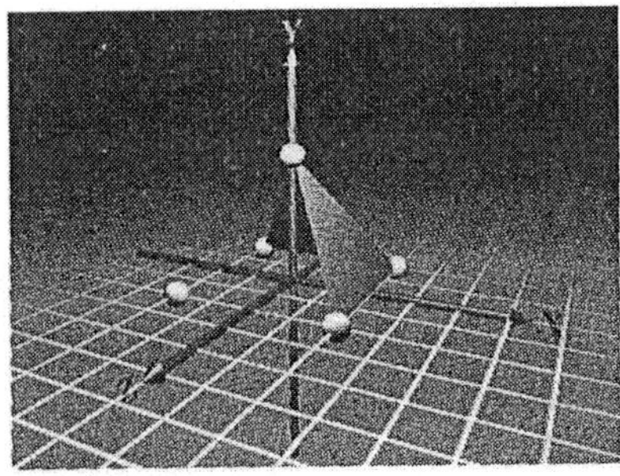

■ axis of rotational symmetry

the line which stays in the same place when we perform a rotation.

■ axis of symmetry

a line that passes through a figure in such a way that the part of the figure on one side of the line is a mirror reflection or image of the part of the figure on the other side of the line.

■ azimuth

1. in astronomy, the measure of how far away from north a star (or anything else) would be if it fell straight down onto the horizon.

2. in 2D polar co-ordinates, the angle between the reference line and the line connecting a point with the origin; in 3D polar co-ordinates, the angle between a reference line, and the line connecting a given point's projection onto the horizontal plane with the origin.

■ Baire category theorem

if X is a complete metric space, then the intersection of a countable collection of dense open subsets of X is also dense. Equivalently, the union of a countable collection of closed, nowhere dense subsets of X is nowhere dense.

■ ball

a sphere together with its interior.

■ Banach space

a complete normed space.

Banach-Tarski Paradox

given two bounded subsets of 3-dimensional Euclidean space, provided each has interior points, then the first may be decomposed into finitely many pieces and translated by rigid motions into a subset congruent to the second subset. For instance, a ball of unit radius can be decomposed into finitely many pieces, and then reassembled into two balls of unit radius. This shocking result is a consequence of the axiom of choice.

band matrix

a matrix whose only nonzero entries lie on a band around the leading diagonal.

band-connected sum

a knot formed from two other knots by connecting them along two parallel segments called a band. Although the original two knots cannot be entangled with each other (they must be separated by a sphere), the band can meander among them in a complicated way. A band-connected sum in which the band is in the simplest possible position is called a connected sum of knots.

bar graph

a diagram using bars whose lengths represent a set of data.

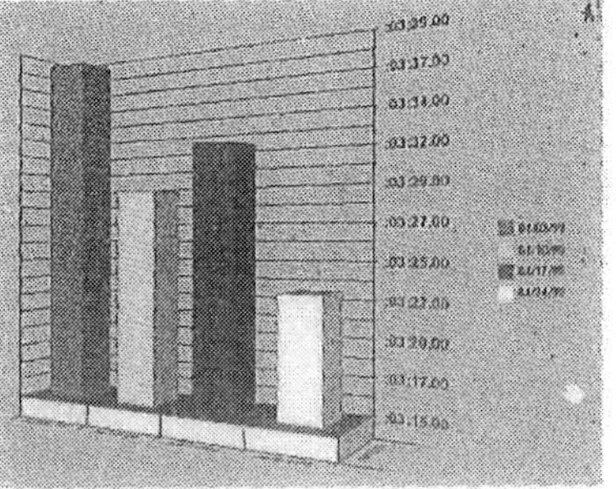

base

in the expression xy, x is called the base and y is the exponent.

base of a percent

the number for which the percent is found. For example, in the problem 10% of 50, the base is 50.

base ten

a number system in which each digit has 10 times the value of the same digit one place to its right. For example, with 77 , the 7 to the left is worth 70, or 10 times the value of the 7 to the right.

basis

in mathematics, usually means basis in the sense of linear algebra; a minimal set of vectors that spans a vector space.

Bayes's Rule

a rule for finding conditional probability.

bead box

a box or dish (clear) with a divider and containing a given number of beads. Used for developing addition and subtraction skill by shaking and getting different combinations.

benchmark

a point of reference used in estimation.

between

point B is between points A and C if AB + BC = AC.

bias

a measurement procedure or estimator is said to be biased if, on the average, it gives an answer that differs from the truth. The bias is the average (expected) difference between the measurement and the truth. For example, if you get on the scale with clothes on, that biases the measurement to be larger than your true weight (this would be a positive bias). The design of an experiment or of a survey can also lead to bias. Bias can be deliberate, but it is not necessarily so.

biconditional statement

a compound statement that says one sentence is true if and only if the other sentence is true.

bijection

a one-to-one onto function.

bilateral symmetry

symmetry about a fixed line in a plane (line of symmetry). Each point P of the symmetrical configuration has an image point P' such that the line of symmetry is the perpendicular bisector of PP'. Points on the line transform into themselves. In three dimensions bilateral symmetry can exist about a fixed plane.

■ **bimodal**
having two modes.

■ **bin**
see class interval.

■ **binary number**
a number written to base 2.

■ **binary number system**
the number system that uses only 0's and 1's. The places in the binary numbers are .... 2^n .... $2^5 = 32$, $2^4 = 16$, $2^3 = 8$, $2^2 = 4$, $2^1 = 2$, $2^0 = 1$.

■ **binary operation**
a binary operation is an operation that involves two operands. For example, addition and subtraction are binary operations.

■ **binomial**
1.an expression that is the sum of two terms.
2. a polynomial that contains only two terms, such as (6x + 2y).

■ **binomial coefficient**
the coefficients of x in the expansion of (x+1)n.

■ **binomial distribution**
a random variable has a binomial distribution (with parameters n and p) if it is the number of "successes" in a fixed number n of independent random trials, all of which have the same probability p of resulting in "success." Under these assumptions, the probability of k successes (and n-k failures) is ${}^nC_k$ pk(1-p)n-k, where ${}^nC_k$ is the number of combinations of n objects taken k at a time ${}^nC_k$ = n!/(k!(n-k)!). The expected value of a random variable with the Binomial distribution is n×p, and the standard error of a random variable with the Binomial distribution is (n×p×(1 - p))½. This page shows the probability histogram of the binomial distribution.

■ **binomial theorem**
the theorem that tells how to expand the expression (a + b)n ... the binomial (a + b) raised to the nth power. There is a descending power of the first term ... an ascending power of the second term ... with numeric values from Pascal's Triangle.

biquadratic equation

a polynomial equation of the 4th degree.

bisect

to cut or divide something in half.

bit

a binary digit.

bivariate

having or having to do with two variables. For example, bivariate data are data where we have two measurements of each "individual." These measurements might be the heights and weights of a group of people (an "individual" is a person), the heights of fathers and sons (an "individual" is a father-son pair), the pressure and temperature of a fixed volume of gas (an "individual" is the volume of gas under a certain set of experimental conditions), etc. Scatterplots, the correlation coefficient, and regression make sense for bivariate data but not univariate data. C.f. univariate.

blind, blind experiment

in a blind experiment, the subjects do not know whether they are in the treatment group or the control group. In order to have a blind experiment with human subjects, it is usually necessary to administer a placebo to the control group.

Bolzano-Weierstrass property

a topological space X is said to have the Bolzano-Weierstrass property if every infinite sequence in X has a convergent subsequence. It is a theorem that every compact space has the Bolzano-Weierstrass property.

bootstrap estimate of standard error

the name for this idea comes from the idiom "to pull oneself up by one's bootstraps," which connotes getting out of a hole without anything to stand on. The idea of the bootstrap is to assume, for the purposes of estimating uncertainties, that the sample is the population, then use the SE for sampling from the sample to estimate the SE of sampling from the population. For sampling from a box of numbers, the SD of the sample is the bootstrap estimate of the SD of the box from which the sample is drawn. For sample percentages, this takes a particularly simple form the SE of the sample percentage of n draws from a box, with replacement, is SD(box)/n½, where for a box that contains only zeros and ones, SD(box) = ((fraction of ones in box)×(fraction of zeros in box))½. The bootstrap estimate of the SE of the sample percentage consists of estimating SD(box) by ((fraction of ones in sample)×(fraction of zeros in sample))½. When the sample size is large, this approximation is likely to be good.

Borel measure

a measure m on a metric space X defined on a σ algebra of sets which contains the Borel sets is called a Borel measure.

Borel set

if X is a topological space, then the σ -algebra of sets generated by the open sets of X is called the Borel σ -algebra, and its members are called Borel sets. The Borel subsets of the real line are generated by the open intervals (equivalently, by the closed intervals, half-open intervals, and rays).

Bourbaki, Nicolas

although the myth of an actual French mathematician named Bourbaki was maintained for many years, it became generally known in 1952 that "Bourbaki" is the nom de plume of a private congress of young French mathematicians which formed in 1935, with the purpose of writing a comprehensive exposition of modern mathematics in many volumes, called Elements. Bourbaki's original members included André Weil, Henri Cartan, Claude

Chevalley, and Jean Dieudonné. Other mathematicians have been recruited into Bourbaki in the years since, including G. de Rham, A. Grothendieck, S. Lang, and R. Thom. A principle motivation for the activity of the Bourbaki group is the felt need to unify the entire corpus of mathematical work in the face of accelerating diversification and specialisation of mathematics.

■ box model

an analogy between an experiment and drawing numbered tickets "at random" from a box with replacement. For example, suppose we are trying to evaluate a cold remedy by giving it or a placebo to a group of n individuals, randomly choosing half the individuals to receive the remedy and half to receive the placebo. Consider the median time to recovery for all the individuals (we assume everyone recovers from the cold eventually; to simplify things, we also assume that no one recovered in exactly the median time, and that n is even). By definition, half the individuals got better in less than the median time, and half in more than the median time. The individuals who received the treatment are a random sample of size n/2 from the set of n subjects, half of whom got better in less than median time, and half in longer than median time. If the remedy is ineffective, the number of subjects who received the remedy and who recovered in less than median time is like the sum of n/2 draws with replacement from a box with two tickets in it one with a "1" on it, and one with a "0" on it. This page illustrates the sampling distribution of random draws with or without from a box of numbered tickets.

■ box-and-whisker plot

a graphical method for showing the median, quartiles, and extremes of data. A box plot shows where the data are spread out and where they are concentrated.

■ **breakdown point**

the breakdown point of an estimator is the smallest fraction of observations one must corrupt to make the estimator take any value one wants.

■ **butterfly effect**

in a system when a small change results in an unpredictable and disproportionate disturbance, the effect causing this is called a butterfly effect.

■ **byte**

the amount of memory needed to represent one character on a computer, typically 8 bits.

■ **calculus**

branch of mathematics concerned with rates of change, gradients of curves, maximum and minimum values of functions, and the calculation of lengths, areas, and volumes. It involves determining areas (integration) and tangents (differentiation) which are mutually inverse.

■ **calculus of variations**

calculus problems, especially differentiation and maximisation, involving functions on a set of functions of a real variable. For example, finding the shape of a cable suspended from both ends.

■ **caliban puzzle**

a logic puzzle in which one is asked to infer one or more facts from a set of given facts.

■ **Cantor set**

a subset of the unit interval (on the real number line) which is a perfect set, nowhere dense, uncountable, and homeomorphic to the entire unit interval. See the article for a complete exposition.

■ **Cantor, Georg (born 1845)**

german mathematician and founder of modern set theory. Cantor's concern with the set-theoretic nature of the real numbers began while he was working on certain properties of trigonometric series during the early 1870's. Inspired by the work of his friend Dedekind, Cantor proved in 1873 that the rational numbers are countable, and later the same year proved that the real numbers themselves are uncountable. These proofs introduced important methods, including the notion of a one-to-one correspondence and the method of diagonalisation. Cantor's results were counterintuitive and broke with established dogma about the nature of infinity. Consequently, he became a controversial figure, opposed by some mathematicians (e.g., Kronecker) who used their influence to stifle Cantor's career as much as possible. This opposition exacerbated Cantor's predisposition to severe depression, and he ultimately died, of a heart attack, in a sanitorium.

■ Cantor's theorem

given any set, a new set may be constructed which is cardinally larger, i.e., whose size is represented by a larger cardinal than the size of the initial set. More specifically, for no set X is there a bijection of X onto its power set. This theorem establishes both that there are different sizes of infinity, and that there is no greatest such "size." Cantor's proof was the first use of a diagonalisation argument to derive a contradiction.

■ Cantor-Bendixson theorem

every closed subset of the real numbers is the disjoint union of a perfect set and a countable set.

■ Cantor-Schröder-bernstein theorem

for any sets A and B, if there exists an injective (one-to-one) function from A into B and also an injective function from B into A, then there exists a surjective (onto) function from A onto B.

capillary wave

a small wave in a body of water whose behaviour is governed by surface tension rather than gravity.

cardinal

a cardinal number represents the size of a set irrespective of the order or structure of its elements. If X is a set, its cardinality is generally indicated by |X|. Two sets A and B are said to have the same cardinality if there is a function from A into B that is bijective (i.e., one-to-one and onto).A cardinal is an initial ordinal, i.e., an ordinal for which there does not exist a bijection onto any lesser ordinal. Thus, all finite ordinals (i.e., the natural numbers) are cardinals, but most transfinite ordinals are not cardinals.

cardinal number

a number that describes how many things in a set.

cardinality

two sets X and Y have the same cardinality if there exists a bijection between them. Specifically, the cardinality of X may be understood as a canonical object meant to represent the proper class of sets to which X is bijective. If the axiom of choice is assumed, then the cardinality of X is the unique cardinal number having the same cardinality as X.

Cartesian coordinates

cartesian coordinates (x,y) specify the position of a point in a plane relative to the horizontal x and the vertical y axes. The x and y axes form the basis of two-dimensional Cartesian coordinate system.

categorical variable

a variable whose value ranges over categories, such as {red, green, blue}, {male, female}, {Arizona, California, Montana, New York}, {short, tall}, {Asian, African-American, Caucasian, Hispanic, Native American, Polynesian}, {straight, curly}, etc. Some categorical variables are ordinal. The distinction between categorical variables and qualitative variables is a bit blurry. C.f. quantitative variable.

category theory

the study of abstracted collections of mathematical objects, such as the category of sets or the category of vector spaces, together with abstracted operations sending one object to another, such as the collection of functions from one set to another or linear transformations from one vector space to another.

catenary

a curve whose equation is $y = (a/2)(ex/a+e-x/a)$. A chain suspended from two points forms this curve.

causality

given an event X in a physical system or a corresponding feature in a partial differential equation, a trichotomy which divides other events into those that may have caused or modified X, those which X can cause or alter, and those which are causally independent from X.

causation, causal relation

two variables are causally related if changes in the value of one cause the other to change. For example, if one heats a rigid container filled with a gas, that causes the pressure of the gas in the container to increase. Two variables can be associated without having any causal relation, and even if two variables have a causal relation, their correlation can be small or zero.

Cayley numbers/ octonions

a non-associative generalisation of the quaternions and the complex numbers involving numbers with one real coefficient and seven imaginary coefficients.

ceiling function

the ceiling function of x is the smallest integer greater than or equal to x.

cellular automaton

a mathematical model consisting of a grid of cells, a notion of neighbouring cells, and a list of states for the cells. The state of each cell at a given iteration of time is a function of its state and that of its neighbours at the previous iteration.

■ **census**

the count of a population.

■ **center**

the point that is the same distance from all the points on a circle. The point that is the same distance from all the points on a sphere. The point inside an ellipse where the major and the minor axes intersect. The center of a circle that can be inscribed in a regular polygon.

■ **central angle**

an angle that has its vertex at the center of a circle.

■ **central limit theorem**

the central limit theorem states that the probability histograms of the sample mean and sample sum of n draws with replacement from a box of labeled tickets converge to a normal curve as the sample size n grows, in the following sense As n grows, the area of the probability histogram for any range of values approaches the area under the normal curve for the same range of values, converted to standard units. See also the normal approximation.

■ **centroid**

the center of mass of an object. The point where the object would balance if supported by a single support. The point in a triangle where the three medians intersect.

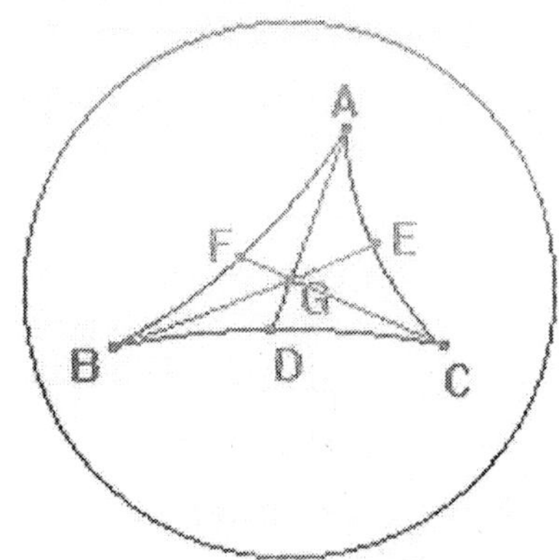

■ **certain event**

an event is certain if its probability is 100%. Even if an event is certain, it might not occur. However, by the

complement rule, the chance that it does not occur is 0%.

cevian

a line segment extending from a vertex of a triangle to the opposite side.

chain rule

dy/dx = dy/du * du/dx.

chance variation, chance error

a random variable can be decomposed into a sum of its expected value and chance variation around its expected value. The expected value of the chance variation is zero; the standard error of the chance variation is the same as the standard error of the random variable—the size of a "typical" difference between the random variable and its expected value. See also sampling error.

change of units or variables

see also transformation.

chaos

apparent randomness whose origins are entirely deterministic. A state of disorder and irregularity whose evolution in time, though governed by simple exact laws, is highly sensitive to starting conditions a small variation in these conditions will produce wildly different results, so that long-term behavior of chaotic systems cannot be predicted. This sensitivity to initial conditions is also known as the butterfly effect (when a butterfly flaps its wings in Mexico, the result may be a hurricane in Florida a month later).

characteristic class

a kind of homological model for a decoration or property of a manifold or other topological space. The simplest characteristic class describes how a manifold fails to be orientable, that is, in which directions a being can travel in the manifold and reverse its handedness.

Chebychev's inequality

for lists For every number k>0, the fraction of elements in a list that are k SD's or further from the arithmetic mean of the list is at most 1/k2. For random variables For every number k>0, the probability that a random variable X is k SEs or further from its expected value is at most 1/k2.

Chern-Simons form

a differential 3-form computed from a connection or gauge field on a manifold. Since it is a 3-form, it is in effect a function that can be naturally integrated over a 3-manifold, and it plays an important role in 3-dimensional topology and quantum field theory.

chi-square curve

the chi-square curve is a family of curves that depend on a parameter called degrees of freedom (d.f.). The chi-square curve is an approximation to the probability histogram of the chi-square statistic for multinomial model if the expected number of outcomes in each category is large. The chi-square curve is positive, and its total area is 100%, so we can think of it as the probability histogram of a random variable. The balance point of the curve is d.f., so the expected value of the corresponding random variable would equal d.f.. The standard error of the corresponding random variable would be (2×d.f.)½. As d.f. grows, the shape of the chi-square curve approaches the shape of the normal curve. This page shows the chi-square curve.

chi-square statistic

the chi-square statistic is used to measure the agreement between categorical data and a multinomial model that predicts the relative frequency of outcomes in each possible category. Suppose there are n independent trials, each of which can result in one of k possible outcomes. Suppose that in each trial, the probability that outcome i occurs is pi, for i = 1, 2, … , k, and that these probabilities are the same in every trial. The expected number of times outcome 1 occurs in the n trials is n×p1; more generally, the ex-

pected number of times outcome i occurs is.
expectedi = n×pi.

■ **chord**
a line segment that connects two points on a curve.

■ **circle**
the set of points in a plane that are a fixed distance from a given point.

■ **circle packing**
an arrangement of round disks in the euclidean-space{Euclidean} or hyperbolic plane or on the round sphere such that no two disks overlap with non-zero area. Depending on the context, the circles may or may not be the same size. A theorem of Koebe, revived by Thurston, states that given any planar graph, there is a circle packing with a circle for each vertex of the graph and kissing circles for each edge.

■ **circular cone**
a cone whose base is a circle.

■ **circular functions**
same as trigonometric functions.

■ **circumcenter**
the point in a triangle that is the center of the circle that can be circumscribed about the triangle. The intersection of the perpendicular bisectors of the triangle.

■ **circumcircle**
the circle circumscribed about a figure.

■ **circumference**
the distance around a closed curve. The circumference of a circle is 2*pi*r where r is the radius of the circle.

■ **circumscribed circle**
a circle that passes through all of the vertices of a regular polygon.

■ **cissoid**
a curve with equation $y^2(a-x)=x^3$.

Cissoid of Diocles

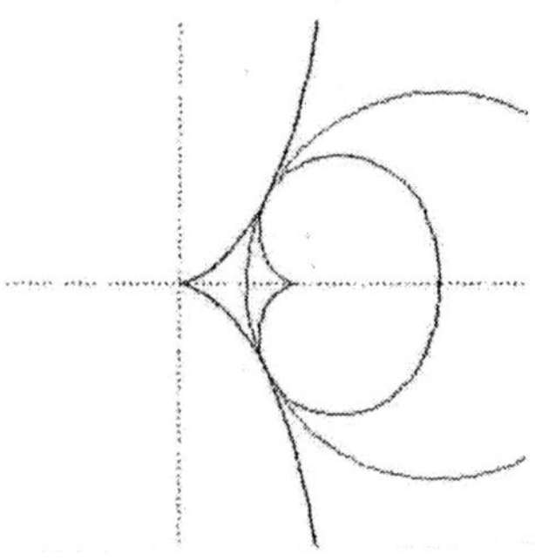
Inverse Curve

class boundary

a point that is the left endpoint of one class interval, and the right endpoint of another class interval.

class interval

in plotting a histogram, one starts by dividing the range of values into a set of non-overlapping intervals, called class intervals, in such a way that every datum is contained in some class interval. See the related entries class boundary and endpoint convention.

classical

in physics and mathematical physics, the term classical sometimes has the narrow meaning of non-quantum; equations of motion interpreted by means of ordinary dynamical systems rather than statistical quantum rules.

classification

the goal in a branch of mathematics of providing an exhaustive list of some type of mathematical object with no repetitions. Example: The classification of 3-manifolds is one of the outstanding problems in topology. With the advent of computers, one weak but precise way to state a classification problem is to ask whether there is an algorithm to determine whether two given objects are equivalent.

closed interval

an interval that contains its endpoints.

closure property

if the result of doing an operation on any two elements of a set is always an element of the set, then the set is closed under the operation. For example, the operations addition and multiplication of natural numbers are closed, but subtraction and division are not.

cluster sample

in a cluster sample, the sampling unit is a collection of population units, not single population units. For example, techniques for adjusting the U.S. census start with a sample of geographic blocks, then (try to) enumerate all inhabitants of the blocks in the sample to obtain a sample of people. This is an example of a cluster sample. (The blocks are chosen separately from different strata, so the overall design is a stratified cluster sample.).

■ clustering

an estimating strategy that involves taking the average amount and multiplying.

■ codimension

in general, if a mathematical object sits inside or is associated to another object of dimension n, then it is said to have codimension k if it has dimension n-k.

■ coefficient

the constant multipliers of the indeterminate variable in a polynomial. For example, in the polynomial x^2+3x+7, the coefficients are 1, 3, and 7.

■ cofunction

the cofunction of a trigonometric function, f(x), is equal to f(pi/2 - x). The cofunction of the sine is the cosine. The cofunction of the secant is the cosecant. The cofunction of the tangent is the cotangent.

■ collinear

points are collinear if they lie on the same line.

■ combinations

the number of combinations of n things taken k at a time is the number of ways of picking a subset of k of the n things, without replacement, and without regard to the order in which the elements of the subset are picked. The number of such combinations is ${}_nC^k = n!/(k!(n-k)!)$, where k! (pronounced "k factorial") is $k\times(k-1)\times(k-2)\times\cdots\times 1$. The numbers ${}_nC^k$ are also called the Binomial coefficients. From a set that has n elements one can form a total of 2n subsets of all sizes. For example, from the set {a, b, c}, which has 3 elements, one can form the 23 = 8 subsets {}, {a}, {b}, {c}, {a,b}, {a,c}, {b,c}, {a,b,c}. Because the number of subsets with k elements one can form from a set with n elements is ${}_nC^k$, and the total number of subsets of a set is the sum of the numbers of possible subsets of each size, it follows that ${}^nC_0 + {}^nC_1 + {}^nC_2 + \dots + {}^nC_n = 2n$. The calculator has a button (${}_nC^m$) that lets you compute the number of combinations of m things chosen from a set of n things. To use the button, first type the value of n, then push the ${}^nC_m$ button, then type the value of m, then press the "=" button.

combinatorial geometry

the visual study of discrete and finite structures, and the study of discrete and finite possibilities for the arrangement or features of geometric objects.

common denominator

a denominator that is common to all the fractions within an equation. The smallest number that is a common multiple of the denominators of two or more fractions is the lowest (or least) common denominator (LCM).

common factor

a whole number that divides exactly into two or more given numbers. The largest common factor for two or more numbers is their highest common factor (HCF).

common logarithm

logarithm with a base of 10 shown as $\log_{10}$ [$\log_{10} 10^x = x$].

common ratio

in a geometric sequence, any term divided by the previous one will give the same common factor which is called the common ratio.

commutative

when numbers can be added or multiplied in any order. e.g., $2 + 3 = 3 + 2$.

commutative property

a binary operation * defined on a set has the commutative property if for every two elements, a and b, $a*b = b*a$. For example, the operations addition and multiplication of natural numbers are commutative, but subtraction and division are not.

■ **compass**

a tool used to draw circles or measure distances.

■ **compatible numbers**

numbers which tend to compute easily.
e.g., 100 - 25, 8 + 2.

■ **compensating**

an estimating strategy that involves adjusting the answer up or down to more closely approximate the value.

■ **complement**

the complement of a subset of a given set is the collection of all elements of the set that are not elements of the subset.

■ **complement rule**

the probability of the complement of an event is 100% minus the probability of the event P(Ac) = 100% - P(A).

■ **complementary angle**

a complementary angle is either of two angles, which add up to 90 degrees.

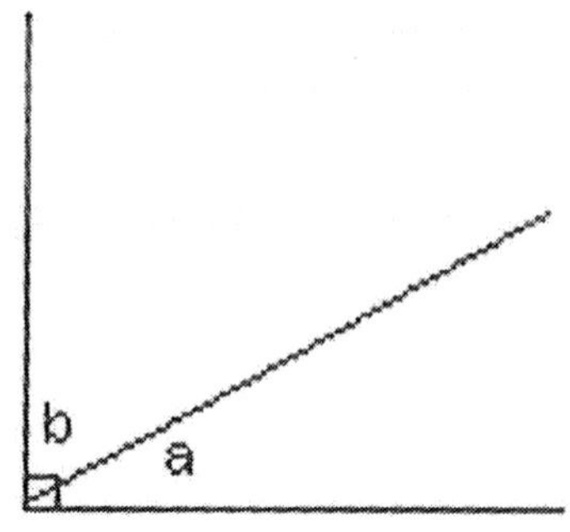

■ **completing the square**

the method of adding an expression to both sides of an equation so that one side becomes a perfect square trinomial.

■ **complex analysis**

the study of functions, especially analytic functions, of one or several complex variables, and related questions concerning Riemann surfaces or complex manifolds.

■ **complex fraction**

a fraction that contains a fraction in its numerator and/or denominator.

■ **complex manifold**

a manifold with complex coordinates; its ordinary or real dimension is then twice its complex dimension.

complex numbers
numbers that have the form a + bi where a and b are real numbers and i satisfies the equation $i^2 = -1$. Multiplication is denoted by $(a+bi)(c+di) = (ac-bd) + (ad+bc)i$, and addition is denoted by $(a+bi) + (c+di) = (a+c) + (b+d)i$.

component
the components in the vector (a, b, c) are a, b, and c.

composite function
a function that consists of two functions arranged in such a way that the output of one function becomes the input of the other function.

composite number
any integer which is not a prime number, i.e., evenly divisible by numbers other than 1 and itself.

compressible fluid
an actual or mathematically abstracted fluid that can change its volume.

computational complexity
the absolute minimum amount of time (or sometimes space) that a computer must take to perform a particular computational task. For example, sorting n numbers has time complexity proportional to n(log n).

computational techniques
operations or tools–number lines, calculators.

concave
a figure is concave if a line segment can be drawn so that it goes in, out, then back into the figure.

CONCAVE

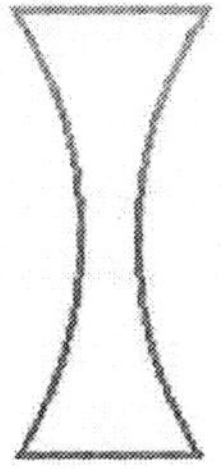

conclusion
the part of an if - then statement that follows the word "then". Consequent.

conditional probability
suppose we are interested in the probability that some event

A occurs, and we learn that the event B occurred. How should we update the probability of A to reflect this new knowledge? This is what the conditional probability does it says how the additional knowledge that B occurred should affect the probability that A occurred quantitatively. For example, suppose that A and B are mutually exclusive. Then if B occurred, A did not, so the conditional probability that A occurred given that B occurred is zero. At the other extreme, suppose that B is a subset of A, so that A must occur whenever B does. Then if we learn that B occurred, A must have occurred too, so the conditional probability that A occurred given that B occurred is 100%. For in-between cases, where A and B intersect, but B is not a subset of A, the conditional probability of A given B is a number between zero and 100%. Basically, one "restricts" the outcome space S to consider only the part of S that is in B, because we know that B occurred. For A to have happened given that B happened requires that AB happened, so we are interested in the event AB. To have a legitimate probability requires that P(S) = 100%, so if we are restricting the outcome space to B, we need to divide by the probability of B to make the probability of this new S be 100%. On this scale, the probability that AB happened is P(AB)/P(B). This is the definition of the conditional probability of A given B, provided P(B) is not zero (division by zero is undefined). Note that the special cases AB = {} (A and B are mutually exclusive) and AB = B (B is a subset of A) agree with our intuition as described at the top of this paragraph. Conditional probabilities satisfy the axioms of probability, just as ordinary probabilities do.

■ conditional statement

an 'if - then' statement.

■ cone

the union of all line segments that connect a point and a

closed curve in a different plane from the point.

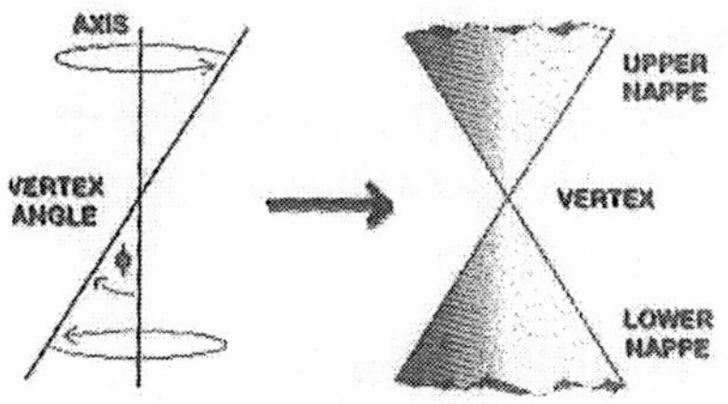

confidence interval

a confidence interval for a parameter is a random interval constructed from data in such a way that the probability that the interval contains the true value of the parameter can be specified before the data are collected.

confidence level

the confidence level of a confidence interval is the chance that the interval that will result once data are collected will contain the corresponding parameter. If one computes confidence intervals again and again from independent data, the long-term limit of the fraction of intervals that contain the parameter is the confidence level.

conformal

angle-preserving or angle-defining. The Mercator map is a conformal map of the Earth because angles are true. A conformal structure on a manifold defines angles between curves segments on the manifold but not their lengths.

confounding

when the differences between the treatment and control groups other than the treatment produce differences in response that are not distinguishable from the effect of the treatment, those differences between the groups are said to be confounded with the effect of the treatment (if any). For example, prominent statisticians questioned whether differences between individuals that led some to smoke and others not to (rather than the act of smoking itself) were responsible for the observed difference in the frequencies with which smokers and non-smokers contract various illnesses. If that were the case, those factors would be confounded with the effect of smoking. Confounding is quite likely to affect observational studies and experiments that are not randomised. Confounding tends to be decreased by randomisation. See also Simpson's Paradox.

■ **congruence**
the state of having the same size and shape.

■ **congruent**
two shapes in the plane or in space are congruent if there is a rigid motion that identifies one with the other (see the definition of rigid motion).

■ **congruent figures**
two geometric figures that are identical in size and shape.

■ **conic section**
1. parabola, hyperbola, ellipse, circle. Formed by the intersection of a plane with a right circular cone.
2.the cross section of a right circular cone cut by a plane. An ellipse, parabola, and hyperbola are conic sections.

■ **conjecture**
a hypothesis drawn from observed patterns in several examples.

■ **conjecturing**
to guess.

■ **conjugate**
the conjugate of a complex number is formed by reversing the sign on the imaginary part of the number. The conjugate of a + bi is a - bi.

■ **connected sum**
1. a manifold formed from two others by removing balls and gluing along the resulting spherical boundary.
2. the analogous operation for knots; a band-connected sum in which the band connects the knots in the simplest possible way (by piercing the separating sphere only once).

■ **consequent**
the part of an "if - then" statement that follows the "then". Conclusion.

conservation law

a Partial differential equation that expresses the fact that some physical quantity is locally conserved in a fluid or other continuous physical system, such as energy, momentum, or the quantity of fluid itself. Typically the behaviour of such a system is completely determined by its conservation laws.

constant

a number which does not change. The number 27 is a constant, as are values like pi and Euler's number.

constant relative width

two convex bodies have constant relative width if the width of one in any given direction is the length of a chord of the other in the same direction.

constant width

a convex body has constant width if any two parallel hyperplanes that touch the convex body on opposite sides are the same distance apart.

contextual situation

relating mathematical problems to real, modeled or illustrated circumstance.

continuity correction

in using the normal approximation to the binomial probability histogram, one can get more accurate answers by finding the area under the normal curve corresponding to half-integers, transformed to standard units. This is clearest if we are seeking the chance of a particular number of successes. For example, suppose we seek to approximate the chance of 10 successes in 25 independent trials, each with probability p = 40% of success. The number of successes in this scenario has a binomial distribution with parameters n = 25 and p = 40%. The expected number of successes is np = 10, and the standard error is $(np(1-p))^{1/2} = 6^{1/2} = 2.45$. If we consider the area under the normal curve at the point 10 successes, transformed to standard units, we get zero the area under a point is always zero. We get a better approximation by considering 10 successes to be the range from 9 1/2 to 10 1/2 successes. The only possible number of successes between 9 1/2 and 10 1/2 is 10, so this is exactly right for the binomial distribution. Because the normal curve is continuous and a binomial random variable is discrete, we need to "smear out" the binomial probability over an appropriate range. The lower endpoint of the range, 9 1/2 successes, is (9.5 - 10)/2.45 = -0.20 standard units. The upper endpoint of the range, 10 1/2 successes, is (10.5 - 10)/2.45 = +0.20 standard units. The area under the normal curve between -0.20 and +0.20 is about 15.8%. The true binomial probability is $25_{C10}\times(0.4)10\times(0.6)15 = 16\%$. In a similar way, if we seek the normal approximation to the probability that a binomial random variable is in the range from i successes to k successes, inclusive, we should find the area under the normal curve from i-1/2 to k+1/2 successes, transformed to standard units. If we seek the probability of more than i successes and fewer than k successes, we should find the area under the normal curve corresponding to the range i+1/2 to k-1/2 successes, transformed to standard units. If we seek the probability of more than i but no more than k successes, we should find the area under the normal curve corresponding to the range i+1/2 to k+1/2 successes, transformed to standard units. If we seek the probability of at least i but fewer than k successes, we should find the area under the normal curve corresponding to the range i-1/2 to k-1/2 successes, transformed to standard units. Including or excluding the half-integer ranges at the ends of the interval in this manner is called the continuity correction.

continuous

a function is continuous if you can draw it without lifting your pencil off the paper. $y = f(x)$ is continuous at a if. 1. $f(a)$ exists. 2. lim as $x \to a$ $f(x)$ exists. And 3. lim as $x \to a$ of $f(x) = f(a)$.

continuous variable

a quantitative variable is continuous if its set of possible values is uncountable. Examples include temperature, exact height, exact age (including parts of a second). In practice, one can never measure a continuous variable to infinite precision, so continuous variables are sometimes approximated by discrete variables. A random variable X is also called continuous if its set of possible values is uncountable, and the chance that it takes any particular value is zero (in symbols, if $P(X = x) = 0$ for every real number x). A random variable is continuous if and only if its cumulative probability distribution function is a continuous function (a function with no jumps).

contrapositive

if p and q are two logical propositions, then the contrapositive of the proposition (p IMPLIES q) is the proposition ((NOT q) IMPLIES (NOT p)). The contrapositive is logically equivalent to the original proposition.

control

in mathematics, a control is a time-dependent function $u(t)$ that influences a dynamical system $dy/dt = F(u,y)$, with the understanding that u should be chosen to minimise some value of the solution $y(t)$ or otherwise optimise some related quantity.

control for a variable

to control for a variable is to try to separate its effect from the treatment effect, so it will not confound with the treatment. There are many methods that try to control for variables. Some are based on matching individuals between treatment and control; others use assumptions about the nature of the effects of the variables to try to model the effect mathematically, for example, using regression.

■ control group

the subjects in a controlled experiment who do not receive the treatment.

■ control theory

the mathematics of programming robots and other machines to respond to changing conditions so that they maintain self-control. For example, the problem of programming an automatic pilot of an airplane so that the plane doesn't crash. See also.

■ controlled experiment

an experiment that uses the method of comparison to evaluate the effect of a treatment by comparing treated subjects with a control group, who do not receive the treatment.

■ controlled, randomised experiment

a controlled experiment in which the assignment of subjects to the treatment group or control group is done at random, for example, by tossing a coin.

■ convenience sample

a sample drawn because of its convenience; not a probability sample. For example, I might take a sample of opinions in Berkeley (where I live) by just asking my 10 nearest neighbors. That would be a sample of convenience, and would be unlikely to be representative of all of Berkeley. Samples of convenience are not typically representative, and it is not typically possible to quantify how unrepresentative results based on samples of convenience will be.

■ converge, convergence

a sequence of numbers x_1, x_2, x_3 ... converges if there is a number x such that for any number E>0, there is a number k (which can depend on

E) such that |xj - x| < E whenever j > k. If such a number x exists, it is called the limit of the sequence x_1, x_2, x_3 … .

■ convergence in probability

a sequence of random variables X_1, X_2, X_3 … converges in probability if there is a random variable X such that for any number E>0, the sequence of numbers.

$p(|X_1 - X| < e)$, $P(|X_2 - X| < e)$, $P(|X_3 - X| < e)$, …

converges to 100%.

■ convergent series

an infinite series that has a finite sum is called convergent.

■ converse

the statement made by interchanging the hypothesis and the conclusion of a statement.

■ converse

if p and q are two logical propositions, then the converse of the proposition (p IMPLIES q) is the proposition (q IMPLIES p).

■ convex

a region in the plane, in Euclidean space, or in some other geometry with lines such as hyperbolic space, is convex if it always contains the line segment connecting two points if it contains the two points themselves. A convex body is, technically, a closed and bounded convex set with non-zero volume.

■ convex geometry

the study of convex shapes, usually in Euclidean space.

■ coordinate

a set of numbers which give the location of a certain point or object.

■ coordinate

A set of numbers that locates the position of a point usually represented by (x,y) values.

■ coordinate axis

the two perpendicular lines that form the x- and y-axis and create four quadrants.

■ coordinate system

1.any set of two or more magnitudes used to locate

points, lines or curves. Commonly placed by using a horizontal axis (x-axis) and vertical axis (y-axis).
2. A rule of correspondence by which two or more quantities locate points unambiguously and which satisfies the further property that points unambiguously determine the quantities; for example, the usual Cartesian coordinates x, y in the plane.

■ **coplanar**

points that lie within the same plane are called coplanar.

■ **coprime**

integers m and n are coprime if $\gcd(m,n)=1$.

■ **corollary**

a statement that can be easily proven once a theorem is proved.

■ **correlation**

1. the relationship between any two random variables which may or may not be independent. It may be expressed in terms of conditional probabilities or the mutual probability distribution of the random variables.
2. the quadratic term in the relationship between two real-valued random variables; the expectation of the product minus the product of the expectations, suitably normalised.

■ **correlation coefficient**

the correlation coefficient r is a measure of how nearly a scatterplot falls on a straight line. The correlation coefficient is always between -1 and +1. To compute the correlation coefficient of a list of pairs of measurements (X,Y), first transform X and Y individually into standard units. Multiply corresponding elements of the transformed pairs to get a single list of numbers. The correlation coefficient is the mean of that list of products. This page contains a tool that lets you generate bivariate data with any correlation coefficient you want.

■ **cosecant**

$\csc x = 1/\sin x$.

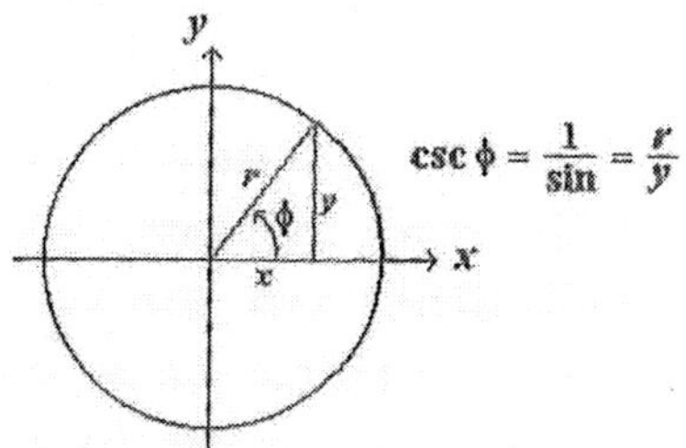

■ **cosine**

the trigonometric function which is defined as the ratio of the leg adjacent to an angle to the hypotenuse of its right triangle.

2. cos(Φ) is the x-coordinate of the point on the unit circle so that the ray connecting the point with the origin makes an angle of Φ with the positive x-axis. When Φ is an angle of a right triangle, then cos (Φ) is the ratio of the adjacent side with the hypotenuse.

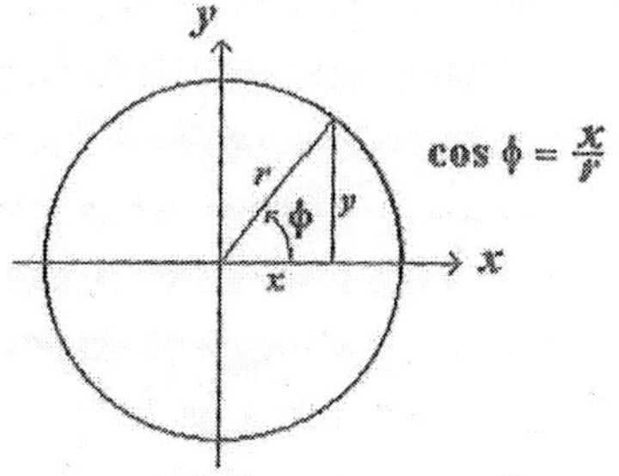

■ **cotangent**

cot x = 1/tan x.

■ **coterminal angles**

angles whose measures are 2kπ apart.

■ **countable set**

a set is countable if its elements can be put in one-to-one correspondence with a subset of the integers. For example, the sets {0, 1, 7, -3}, {red, green, blue}, { … ,-2, -1, 0, 1, 2, . . . }, {straight, curly}, and the set of all fractions, are countable. If a set is not countable, it is uncountable. The set of all real numbers is uncountable.

■ **counter example**

an example of a conditional statement in which the hypothesis is true and the conclusion is false.

■ **counting number**

an element of the set C = {1,2,3,...}.

■ **cover**

a confidence interval is said to cover if the interval contains the true value of the parameter. Before the data are collected, the chance that the confidence interval will contain the parameter value is the coverage probability, which equals the confidence level after the data are collected and the confidence interval is actually computed.

■ coverage probability
the coverage probability of a procedure for making confidence intervals is the chance that the procedure produces an interval that covers the truth.

■ critical point
the point on a curve where the first derivative equals zero. Extremum.

■ critical value
the critical value in an hypothesis test is the value of the test statistic beyond which we would reject the null hypothesis. The critical value is set so that the probability that the test statistic is beyond the critical value is at most equal to the significance level if the null hypothesis be true.

■ cross-sectional study
a cross-sectional study compares different individuals to each other at the same time—it looks at a cross-section of a population. The differences between those individuals can confound with the effect being explored. For example, in trying to determine the effect of age on sexual promiscuity, a cross-sectional study would be likely to confound the effect of age with the effect of the mores the subjects were taught as children the older individuals were probably raised with a very different attitude towards promiscuity than the younger subjects. Thus it would be imprudent to attribute differences in promiscuity to the aging process. C.f. longitudinal study.

■ cryptarithm
a number puzzle in which an indicated arithmetical operation has some or all of its digits replaced by letters or symbols and where the restoration of the original digits is required. Each letter represents a unique digit.

cube
a solid figure with six square faces.

cube root
the factor of a number that, when it is cubed (i.e., x3) gives that number.

cubic
a polynomial of degree 3.

cubic equation
a polynomial equation of degree 3.

cumulative probability distribution function (cdf)
the cumulative distribution function of a random variable is the chance that the random variable is less than or equal to x, as a function of x. In symbols, if F is the cdf of the random variable X, then F(x) = P(X <= x). The cumulative distribution function must tend to zero as x approaches minus infinity, and must tend to unity as x approaches infinity. It is a positive function, and increases monotonically if y > x, then F(y) >= F(x). The cumulative distribution function completely characterises the probability distribution of a random variable.

curvature
in Riemannian geometry, usually means the intrinsic curvature of a manifold with a Riemannian metric. The curvature at a point is positive (negative) if the sum of the angles of a small approximate triangle at that point is greater than (less than) 180 degrees.

curve
a line that is continuously bent.

curve fitting
plotting data and observing the pattern to predict trends.

cusp
a trumpet-shaped salient of a hyperbolic structure or Riemannian metric which is typically infinitely long. The neighbourhood of an ideal vertex is a kind of cusp.

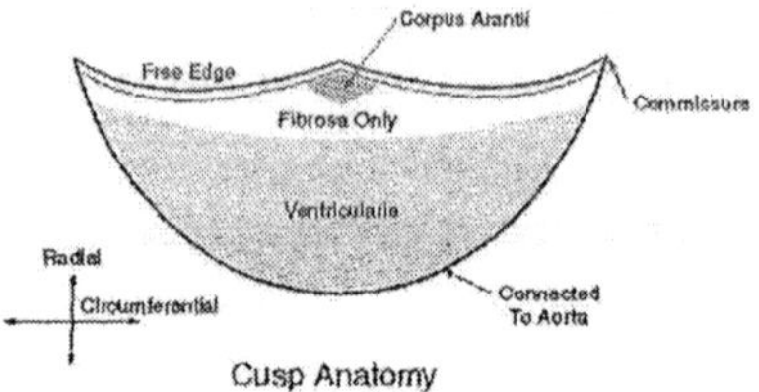

Cusp Anatomy

■ **cyclically symmetric, self-complementary plane partition/ CSSCPP**

a plane partition in a cube which is symmetric under sending the x axis to the y axis, y to z, and z to x, and at the same time is equal to its complement; the set of cubes in the box not in the plane partition considered again as a plane partition by inverting all three axes. Equivalently, a lozenge tiling of a regular hexagon invariant under 60 degree rotation.

■ **cylinder**

a roller shaped figure.
e.g., tin can.

■ **cylinder**

the union of all line segments that connect corresponding points on congruent circles in parallel planes.

■ **data**

information, often numerical, that is usually organized in graphs or charts to show relationships.

■ **decagon**

a polygon with 10 sides.

■ **decimal**

a fraction having a power of ten as denominator, such as 0.34 = 34/100 (10^2) or 0.344 = 344/1000 (10^3). In the continent, a comma is used as the decimal point (between the unit figure and the numerator).

■ **decimal (decimal number)**

a number written using base ten; a number containing a decimal point.

decimal point
a dot separating the ones and the tenths places in a decimal number.

decision problem
the problem of producing a computational algorithm that answers some yes/no question in a finite amount of time. For example, the question of whether two given knots are equivalent, or the question of whether an algebraic equation has a solution in the rational numbers.

deck of cards
a standard deck of playing cards contains 52 cards, 13 each of four suits spades, hearts, diamonds, and clubs. The thirteen cards of each suit are {ace, 2, 3, 4, 5, 6, 7, 8, 9, 10, jack, queen, king}. The face cards are {jack, queen, king}. It is typically assumed that if a deck of cards is shuffled well, it is equally likely to be in each possible ordering. There are 52! (52 factorial) possible orderings.

decreasing function
a function is considered to be decreasing if $f(b) < f(a)$ when $b > a$.

deduction
a conclusion arrived at by reasoning.

deductive reasoning
a series of logical steps in which a conclusion is drawn directly from a set of statements that are known or assumed to be true.

deficient number
a positive integer that is larger than the sum of its proper divisors.

definite integral
the definite integral of $f(x)$ between a and b represents the area under the curve $y = f(x)$, above the x - axis, to the right of the line $x = a$, and to.

the left of the line $x = b$. The definite integral of $f(x) = F(b) - F(a)$ where F is an antiderivative function for $f(x)$.

degenerate diffusion
a diffusion process that only allows diffusion in pre speci-

fied directions instead of in every direction.

degree

1. 1/360 of a full rotation. There are 360 degrees in a circle. Unit of measure of an angle.

2.the degree of a term in one variable is the exponent of that variable. For example, the degree of 7x5 is 5.

Dehn filling

another view of Dehn surgery due to Thurston and motivated by hyperbolic geometry. If the complement of a knot in a 3-manifold has a family of hyperbolic structures, then typically many of them can be completed or filled in to realise different Dehn surgeries along the knot.

Dehn surgery

a modification of a 3-manifold in which a solid torus around a knot is removed and replaced by a solid torus in a different position.

denominator

the number that represents the number of equal parts into which the whole has been divided. In ²/3 the denominator is 3 (lower number in a fraction).

density, density scale

the vertical axis of a histogram has units of percent per unit of the horizontal axis. This is called a density scale; it measures how "dense" the observations are in each bin. See also probability density.

dependent events, dependent random variables

two events or random variables are dependent if they are not independent.

dependent variable

in regression, the variable whose values are supposed to be explained by changes in the other variable (the the independent or explanatory variable). Usually one regresses the dependent variable on the independent variable.

derivative

the derivative at a point on a curve is the gradient of the tangent to the curve at the given point. More technically, a function (f'(x0)) of a function y = f(x), representing the rate of change of y and the gradient of the graph at the point where x = x0, usually shown as dy/dx. The notation dy/dx suggests the ratio of two numbers dy and dx (denoting infinitesimal changes in y and x), but it is a single number, the limit of a ratio (k/h) as they both approach zero. Differentiation is the process of calculating derivatives. The derivatives of all commonly occurring functions are known.

Descartes, René

French mathematician who is generally considered to have laid the foundations for modern mathematics. His greatest achievement was the invention of analytic geometry, in which the methods of algebra and those of geometry are used together. He is also a central figure in the history of modern philosophy; his treatises Meditations and Discourse on Method laid the groundwork both for modern rationalism and modern skepticism. Descartes did not actually use the rectangular coordinate system known as the Cartesian plane (this was developed by Leibniz and others), and he permitted only positive values for his variables. Nonetheless, his development of algebraic methods in geometry made possible an explosion of analytic discoveries by his successors, the most important of which was the discovery of the calculus less than a generation after Descartes' death.

deviation

a deviation is the difference between a datum and some reference value, typically the mean of the data. In computing the SD, one finds the rms of the deviations from the mean, the differences between the individual data and the mean of the data.

diagonal

in a polygon, the line segment joining a vertex with another (non-adjacent) vertex is called a diagonal.

diameter

1.the longest chord of a figure. In a circle, a diameter is a chord that passes through the centre of the circle.

2. any line which goes from one side of a circle to the other, passing directly through the centre point (splits the circle into halves).

diffeomorphism

a bijection between two manifolds that preserves all smooth structure.

difference

the amount that remains after one quantity is subtracted from another.

differential calculus

Differentiation is concerned with rates of change and calculating the gradient at any point from the equation of the curve, y = f(x).

differential equation

Equations involving total or partial differentiation coefficients and, about the rate of change; the difference between some quantity now and its value an instant into the future.

differential geometry

the general study of smooth manifolds decorated by continuous structures such as foliations, Riemannian metrics, and symplectic structures. Riemannian geometry is a disproportionate part of differential geometry.

diffusion equation

a partial differential equation that models the statistics or distribution of many particles undergoing Brownian motion, or the diffusion of one fluid in another fluid, or the diffusion of heat.

digimetic

a cryptarithm in which digits represent other digits.

digit

any one of following ten symbols {0, 1, 2, 3, 4, 5, 6, 7, 8, 9}.

dihedral angle

the angle formed by two planes meeting in space.

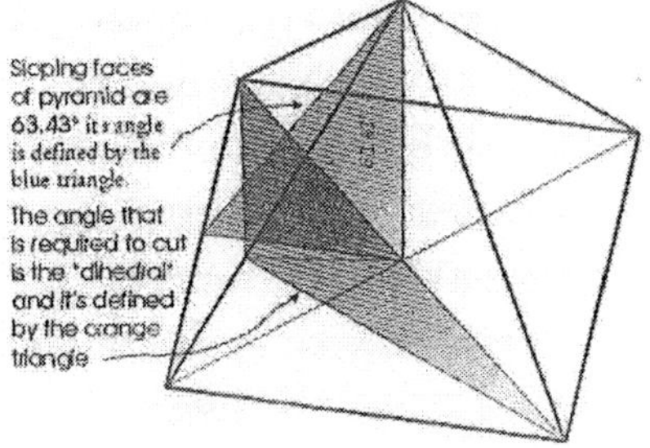

■ **dilation**

in geometry, a transformation D of the plane or space is a dilation at a point P if it takes P to itself, preserves angles, multiplies distances from P by a positive real number r, and takes every ray through P onto itself. In case P is the origin for a Cartesian coordinate system in the plane, then the dilation D maps the point (x, y) to the point (rx, ry).

■ **dimension**

either the length and/or width of a flat surface (two-dimensional); or the length, width, and/or height of a solid (three-dimensional).

■ **dimensional analysis**

a tool for obtaining information about physical systems too complicated for full mathematical solutions to be feasible. It enables one to predict the behavior of large systems from a study of small scale models. A convenient means of checking mathematical equations.

■ **dimensional analysis**

a method of manipulating unit measures algebraically to determine the proper units for a quantity computed algebraically. For example, velocity has units of the form length over time (e.g., meters per second [m/sec]), and acceleration has units of velocity over time; so it follows that acceleration has units (m/sec)/sec = m/(sec2).

■ **diophantine equation**

an equation that is to be solved in integers.

■ **direct proof**

a conclusion proved through deductive reasoning.

■ **directed graphs**

usually equivalent to an arrow diagram, a set of points or nodes connected by lines or arcs, each of which is associated with a given direction.

directly proportional

y is directly proportional to x if y = kx.

disc

a circle together with its interior.

discrete math

the study of mathematical properties of sets and systems that have only a specific number of elements. For example, the results of tossing dice form a discrete set of events, since a die has to land on one of its six faces.

discrete variable

a quantitative variable whose set of possible values is countable. Typical examples of discrete variables are variables whose possible values are a subset of the integers, such as Social Security numbers, the number of people in a family, ages rounded to the nearest year, etc. Discrete variables are "chunky." C.f. continuous variable. A discrete random variable is one whose set of possible values is countable. A random variable is discrete if and only if its cumulative probability distribution function is a stair-step function; i.e., if it is piecewise constant and only increases by jumps.

discriminant

the discriminant of a quadratic equation, $ax^2 + bx + c = 0$ is $b^2 - 4ac$. The discriminant tells how many roots there are for the equation and the nature of the roots.

disjoint or mutually exclusive events

two events are disjoint or mutually exclusive if the occurrence of one is incompatible with the occurrence of the other; that is, if they can't both happen at once (if they have no outcome in common). Equivalently, two events are disjoint if their intersection is the empty set.

disjoint or mutually exclusive sets

two sets are disjoint or mutually exclusive if they have no element in common. Equivalently, two sets are disjoint if their intersection is the empty set.

■ **disjunction**

an OR statement. (concept of set union.).

■ **distribution**

the distribution of a set of numerical data is how their values are distributed over the real numbers. It is completely characterised by the empirical distribution function. Similarly, the probability distribution of a random variable is completely characterised by its probability distribution function. Sometimes the word "distribution" is used as a synonym for the empirical distribution function or the probability distribution function.

■ **distribution function, empirical**

the empirical (cumulative) distribution function of a set of numerical data is, for each real value of x, the fraction of observations that are less than or equal to x. A plot of the empirical distribution function is an uneven set of stairs. The width of the stairs is the spacing between adjacent data; the height of the stairs depends on how many data have exactly the same value. The distribution function is zero for small enough (negative) values of x, and is unity for large enough values of x. It increases monotonically if $y > x$, the empirical distribution function evaluated at y is at least as large as the empirical distribution function evaluated at x.

■ **distribution/probability distribution**

the defining description of a random variable; the set of all possible values of a random variable together with the probability of attaining each value or the probability that the value lands in any given range.

■ **distributive**

a product can be written as the sum of two products.

e.g., 2 x (4 + 5) = 2 x 4 + 2 x 5

3 x 12 = 3 x 10 + 3 x 2

4 x 9 = 4 x 10 - 4 x 1

■ **distributive law**

the formula a(x+y)=ax+ay. distributive property.

a binary operation * is distributive over another binary operation ^ if, a*(b^c) = (a*b)^(a*c). For example, the operation of multiplication is distributive over the operations of addition and subtraction in the set of natural numbers.

■ **divergent series**

a series whose sum is infinite.

■ **dividend**

1.in the expression "a divided by b", a is the dividend and b is the divisor.

2.the number in a division problem which is actually being divided. In 4/3, four is the dividend.

■ **division**

the operation of ascertaining how many times one number, the divisor, is contained in another, the dividend. The result is the quotient, and any number left over is called the remainder. The dividend and divisor are also called the numerator and denominator, respectively.

■ **divisor**

1. in the expression "a divided by b", a is the divident and b is the divisor.

2. the nonzero integer d is a divisor of the integer n if n/d is an integer.

■ **dodecagon**

a polygon with 12 sides.

■ **dodecahedron**

a solid figure with 12 faces A regular dodecahedron is a regular polyhedron with 12 faces. Each face is a rgular pentagon.

■ **Dodgson, Charles Lutwidge (born 1832)**

Oxford mathematician most famous for the books Alice in Wonderland and Through the Looking Glass, which he wrote under the pseudonym Lewis Carroll. However, he also wrote several mathematics textbooks, and delighted in inventing bizarre and humorous syllogisms for exercises in Aristotelian logic. These syllogisms commonly appear in introductory texts on logic to this day. He also formulated what is now called Carroll's Paradox, which shows the need for formal rules of inference in any system of logic.

■ domain

the domain of a function f(x) is the set of x values for which the function is defined.

■ Donaldson theory/ Seiberg-Witten theory

a theory invented by Donaldson and revolutionised by Seiberg and Witten that derives many topological properties of smooth 4-manifolds using gauge theory.

■ double-blind, double-blind experiment

in a double-blind experiment, neither the subjects nor the people evaluating the subjects knows who is in the treatment group and who is in the control group. This mitigates the placebo effect and guards against conscious and unconscious prejudice for or against the treatment on the part of the evaluators.

■ Dugundji extension theorem

a generalisation of the Tietze extension theorem to functions taking values in a Banach space. The gist of the result is that if X is closed in a metrisable Y, then the extension of a real valued function f on X to Y can be linear and continuous as a function of f.

■ duodecimal number system

the system of numeration with base 12.

■ dynamical system

a flow on a space or an iterative function from a space to itself. By Platonist convention and because of scientific applications, a dynamical system is said to move the points of a space around over time, and dynamical systems are studied to determine trajectories under this motion.

■ dynamics

the branch of mathematics which studies the way in

which force produces motion.

e

1. 2.718281828..... The base of the natural logarithm function. e can be found from the series 2 + 1/2! + 1/3! + 1/4! + 1/5! + ...
2.Euler's number, equal to about 2.718, used as the base in a natural logarithm.
3. symbol for the base of natural logarithms (2.7182818285...), defined as the limiting value of (1 + 1/m)m.

earthquake

a discontinuous, surjective map from the hyperbolic plane to itself which moves different pieces isometricaly, and the pieces are separated by hyperbolic lines that make a lamination. A homeomorphism of the circle at infinity of the hyperbolic plane to itself, if it is sufficiently nice in a natural sense, extends uniquely to an earthquake.

eccentricity

a number that indicates the shape of a conic section. The eccentricity of an ellipse is given by e = sqrt(a^2 - b^2) / a.

ecological correlation

the correlation between averages of groups of individuals, instead of individuals. Ecological correlation can be misleading about the association of individuals.

Egyptian fraction

a number of the form 1/x where x is an integer is called an Egyptian fraction.

eigenvalue

characteristic value.

element

a member of a set.

elementary function

one of the functions: rational functions, trigonometric functions, exponential functions, and logarithmic functions.

ellipse

1. a plane figure whose equation is $\frac{x^2}{a^2} + \frac{y^2}{b^2} = 1$

2. a curved line where the sum of the distances from two points (foci) to each point on the curve is constant.

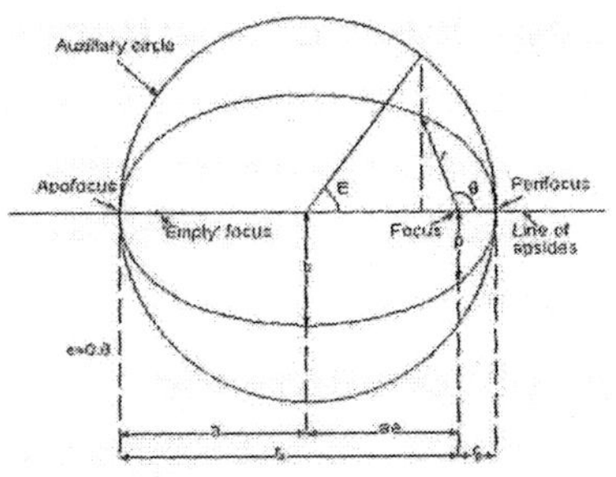

■ **ellipsoid**

a solid figure whose equation is $\frac{x^2}{a^2} + \frac{y^2}{b^2} + \frac{z^2}{c^2} = 1$.

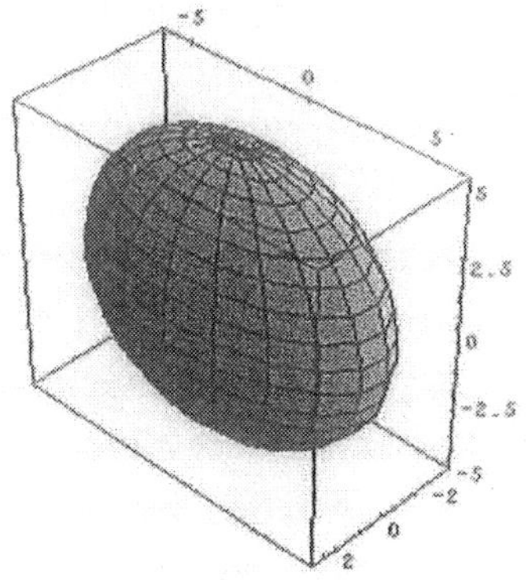

■ **empirical**

relating to the collection of actual data.

■ **empirical law of averages**

the Empirical Law of Averages lies at the base of the frequency theory of probability. This law, which is, in fact, an assumption about how the world works, rather than a mathematical or physical law, states that if one repeats a random experiment over and over, independently and under "identical" conditions, the fraction of trials that result in a given outcome converges to a limit as the number of trials grows without bound.

■ **empty set**

a set that contains no elements.

■ **endpoint convention**

in plotting a histogram, one must decide whether to include a datum that lies at a class boundary with the class interval to the left or the right of the boundary. The rule for making this assignment is called an endpoint convention. The two standard endpoint conventions are (1) to include the left endpoint of all class intervals and exclude the right, except for the rightmost class interval, which includes both of its endpoints, and (2) to include the right endpoint of all class intervals and exclude the left, except for the leftmost interval, which includes both of its endpoints.

■ **ensemble**

in physics, a probability distribution on the states of a complicated system being studied. For example, the thermal probability distribution of a physical object is called the canonical ensemble.

■ **enumerable set**

a countable set.

■ **enumeration**

the goal in combinatorics of finding a simple formula for the number of some type of mathematical object. For example, there are n! permutations of n distinct letters

■ **enumerative combinatorics**

the branch of mathematics concerned with finite enumeration or counting problems.

■ **epi/hypo-convergence**

a more complicated version of epi-convergence suitable for sequences of saddle functions.

■ **epi-convergence**

a notion of convergence of a sequence of real-valued functions useful in minimisation problems and the calculus of variations. More technically, a sequence of functions F_1, F_2, ... epi-converges to F at a point x if some subsequence of some F_1(x_1), F_2(x_2), ... converges to F(x), but no subseqence of any F_1(x_1), F_2(x_2), ... converges to a number smaller than F(x), where in both cases x_1, x_2,... converges to x.

■ **equation**

an algebraic expression in which two values are shown as being equal: 4 = 2+2

■ **equiangular polygon**

a polygon all of whose interior angles are equal.

equichordal point
a point inside a closed convex curve in the plane is called an equichordal point if all chords through that point have the same length.

equilateral
a figure that has the same measurement for each side.

equilateral polygon
a polygon all of whose sides are equal.

equilateral triangle
a triangle that has three equal sides.

equivalent
two expressions that name the same number. For example, 4.6 and 4.60 are equivalent decimals; 2/3 and 4/6 are equivalent fractions; and 2,6 and 1,3 are equivalent ratios.

equivalent equations
two equations whose solutions are the same. For example.
$x + 3y = 10$, and $2x + 6y = 20$.

escribed circle
an escribed circle of a triangle is a circle tangent to one side of the triangle and to the extensions of the other sides.

essential
in low-dimensional topology, a non-trivial kind of curve or surface in a manifold that only sometimes exists. An essential sphere in a 3-manifold, for example, is a sphere that does not bound a ball.

estimate
an answer that has been approximated to be near the actual answer, used to save time when accuracy is not important.

estimator
an estimator is a rule for "guessing" the value of a population parameter based on a random sample from the population. An estimator is a random variable, because its value depends on which particular sample is obtained, which is random. A canonical example of an

estimator is the sample mean, which is an estimator of the population mean.

Euclidean space

a finite-dimensional vector space, such as ordinary 3-dimensional space, together with a metric that satisfies the Pythagorean theorem.

Euclidean transformations

in geometry, the process of changing one configuration into another, including slides, rotations and reflections.

Euler characteristic

an integer associated to a manifold or other topological space which is particularly easy to compute. Example: the Euler characteristic of a surface is given by the number of faces minus edges plus vertices.

Euler line

the Euler line of a triangle is the line connecting the centroid and the circumcentre.

Euler relation

an identity that may be satisfied by a function f on convex polytopes in Euclidean space. The positive Euler relation says that which f(C) is the sum of its values for all faces of C of all dimensions; the negative relation says that it is the alternating sum.

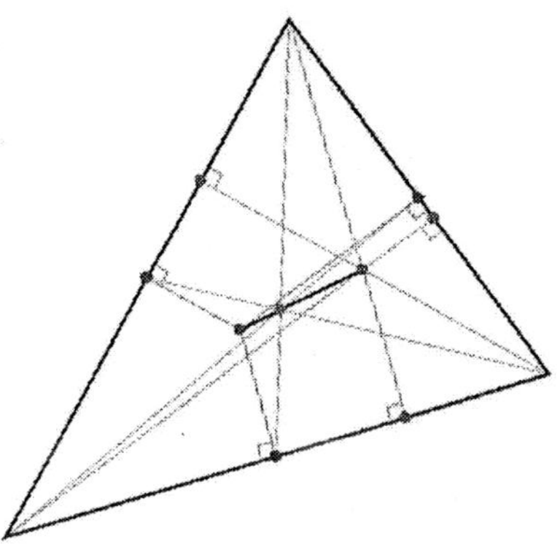

Euler's constant

the limit of the series 1/1+1/2+1/3+...+1/n-ln n as n goes to infinity. Its value is approximately 0.577216.

even function

a function that satisfies the property that f(x) = f(-x).

even number

an integer that is divisible by 2.

event

an event is a subset of outcome space. An event determined by a random variable is an event of the form A=(X is in A). When the random variable X is observed, that determines whether or not A occurs if the value of X happens to be in A, A occurs; if not, A does not occur.

exactly solvable

a statistical mechanical model is called exactly solvable if its main basic parameters, such as its entropy, have simple foı mulas or can be quickly calculated. It is does not mean that the model is completely understood.

excentre

the centre of an excircle.

excircle

an escribed circle of a triangle.

exhaustive

a collection of events {A1, A2, A3, … } is exhaustive if at least one of them must occur; that is, if.

s = A1 U A2 U A3 U . . .

where S is the outcome space.

expanded form

the form a quantity takes when written as a sum of terms, or as a continued product, or in general in any type of extended form. The writing of an expression in full.

(e.g., $(x + a)^3 = x^3 + 3 x 2a + 3 x a^2 + a^3$).

expectation, expected value

the expected value of a random variable is the long-term limiting average of its values in independent repeated experiments. The expected value of the random variable X is denoted EX or E(X). For a discrete random variable (one that has a countable number of possible values) the expected value is the weighted average of its possible values, where the weight assigned to each possible value is the chance that the random variable takes that value. One can think of the expected value of a random variable as the point at which its probability histogram would balance, if it were cut out of a uniform material. Taking the expected value is a linear operation if X and Y are two random variables, the expected value of their sum is the sum of their expected values (E(X+Y) = E(X) + E(Y)), and the expected value of a constant a times a random variable X is the constant times the expected value of X (E(a×X) = a× E(X)).

■ experiment

what distinguishes an experiment from an observational study is that in an experiment, the experimenter decides who receives the treatment.

■ explanatory variable

in regression, the explanatory or independent variable is the one that is supposed to "explain" the other. For example, in examining crop yield versus quantity of fertilizer applied, the quantity of fertilizer would be the explanatory or independent variable, and the crop yield would be the dependent variable. In experiments, the explanatory variable is the one that is manipulated; the one that is observed is the dependent variable.

■ explicit function

an explicit function of x is a function whose values are given by an explicit expression (algebraic or otherwise) in x. For example, the equation y=2x-3 gives values of y as an explicit function of x (solving for x in terms of y would express the value of x as an explicit function of y).

■ expoential function to base a

the function $f(x)=a^x$.

■ exponent

1. in the expression xy, x is called the base and y is called the exponent.

2. a number representing how many times to multiply a number by itself. 23 = 2 x 2 x 2 = 8

■ **exponent power**

A number denoted by a small numeral placed above and to the right of a numerical quantity, which indicates the number of times that quantity is multiplied by itself. In the case of Xn, it is said that X is raised to the power of n.

■ **exponential function**

a function in the form of f(x) = ax where x is a real number, and a is positive and not 1. The most important exponential function is f(x) = ex.

■ **exponential growth**

a group, geometric space, or other mathematical object is said to have exponential growth if its points or elements have some kind of measure of complexity such as length and the number of elements with complexity n exponentiates as n grows. For example, the set of words pronouncable by an English speaker has exponential growth in the length of the word.

■ **expression**

a mathematical phrase with no equal sign, such as 3x, 6, 2n + 3m.

■ **exradius**

an ex-radius of a triangle is the radius of an escribed circle.

■ **extrapolate**

when estimating from a graph, the estimation of a value beyond the given points.

■ **extremum**

a point where a function reaches a maximum or a minimum.

■ **face angle**

the plane angle formed by adjacent edges of a polygonal angle in space.

■ **factor**

when two or more natural numbers are multiplied, each of the numbers is a factor of the product. A factor is then a number by which another number is exactly divided (a divisor).

factor theorem

if $P(x)$ is a polynomial, then if $P(r) = 0$, then $(x - r)$ is a factor of $P(x)$.

factorial

for an integer k that is greater than or equal to 1, k! (pronounced "k factorial") is $k\times(k-1)\times(k-2)\times \ldots \times 1$. By convention, $0! = 1$. There are k! ways of ordering k distinct objects. For example, 9! is the number of batting orders of 9 baseball players, and 52! is the number of different ways a standard deck of playing cards can be ordered. The calculator above has a button to compute the factorial of a number. To compute k!, first type the value of k, then press the button labeled "!".

factorisation

writing a number as the product of its factors which are prime numbers.

fair bet

a fair bet is one for which the expected value of the payoff is zero, after accounting for the cost of the bet. For example, suppose I offer to pay you \$2 if a fair coin lands heads, but you must ante up \$1 to play. Your expected payoff is -\$1+ \$0×P(tails) + \$2×P(heads) = -\$1 + \$2×50% = \$0. This is a fair bet—in the long run, if you made this bet over and over again, you would expect to break even.

false discovery rate

in testing a collection of hypotheses, the false discovery rate is the fraction of rejected null hypotheses that are rejected erroneously (the number of Type I errors divided by the number of rejected null hypotheses), with the convention that if no hypothesis is rejected, the false discovery rate is zero.

Farey sequence

the sequence obtained by arranging in numerical order all the proper fractions having denominators not greater than a given integer.

Fermat number

a number of the form $2\,(2^n) + 1$.

Fermat prime
any prime number in the form of $22n + 1$ (see also Mersenne prime).

Fermat's little theorem
if p is a prime number and b is any whole number, then bp-b is a multiple of p (23 - 2 = 6 and is divisible by 3).

Fermat's spiral
a parabolic spiral.

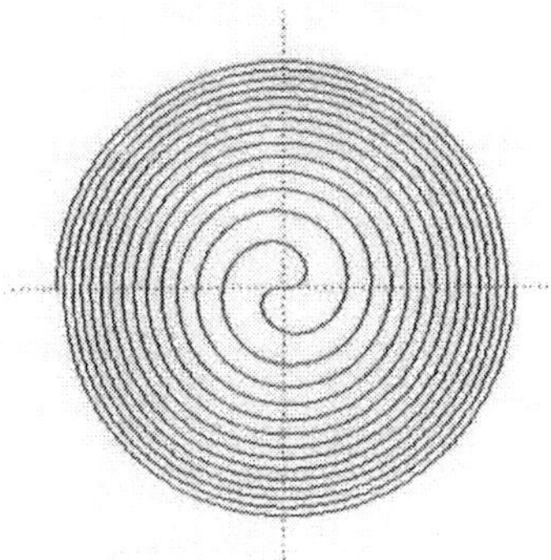

fiber bundle
a topological space that divides into locally parallel fibers that may twist globally. The set of fibers is itself a topological space called the base space. For example, a Möbius strip is a fiber bundle over a circle with line segments as fibers.

Fibonacci Sequence
1, 1, 2, 3, 5, 8, 13, 21, 34, 55, 89, 144, 233, 377.... The sequence in which every number is the sum of the two preceding numbers.

Fibonacci(born 1170)
medieval Italian mathematician, whose real name was Leonardo of Pisa. ("Fibonacci" is a nickname meaning "son of good nature.") Although best known for the number-sequence which bears his name, his most important contribution to mathematics was introducing the Arabic numerals to Europe through his book Liber Abaci (pub. 1202), a treatise on algebraic meth-

ods of arithmetic, and the application of these methods in business. Fibonacci was a businessman and the son of a businessman, educated in North Africa, and it was on his extensive travels in the near east and Africa that he became familiar with the notation which made practical commerce so much easier, since it was a place notation and hence made algorithmic arithmetic calculations possible. It was in the Liber Abaci that the problem which leads to his famous sequence was posed and solved:

■ **fields medal**

by convention, the most prestigious award for research in mathematics. It is awarded every four years to between two and four mathematicians under the age of 40.

■ **figurate numbers**

polygonal numbers.

■ **finite graph**

a structure consisting of vertices and edges, where the edges indicate a mapping among the vertices (e.g., the vertices may represent players in a tournament, and the edges indicate who plays whom).

■ **finite group**

a group containing a finite number of elements.

■ **finite population correction**

when sampling without replacement, as in a simple random sample, the SE of sample sums and sample means depends on the fraction of the population that is in the sample the greater the fraction, the smaller the SE. Sampling with replacement is like sampling from an infinitely large population. The adjustment to the SE for sampling without replacement is called the finite population correction. The SE for sampling without replacement is smaller than the SE for sampling with replacement by the finite population correction factor $((N - n)/(N - 1))^{1/2}$. Note that for sample size n=1, there is no difference between sampling with and without replacement; the finite population correction is then unity. If the sample size is the entire population of N units, there is no variability in the result of sampling without replacement (every member of the population is in the sample exactly once), and the SE should be zero. This is indeed what the finite population correction gives (the numerator vanishes).

finite sequence

a sequence with a finite number of terms. The number of terms in a finite sequence can be counted, and in every finite sequence there is a last term.

Fisher's exact test (for the equality of two percentages)

consider two populations of zeros and ones. Let p1 be the proportion of ones in the first population, and let p2 be the proportion of ones in the second population. We would like to test the null hypothesis that p1 = p2 on the basis of a simple random sample from each population. Let n1 be the size of the sample from population 1, and let n2 be the size of the sample from population 2. Let G be the total number of ones in both samples. If the null hypothesis be true, the two samples are like one larger sample from a single population of zeros and ones. The allocation of ones between the two samples would be expected to be proportional to the relative sizes of the samples, but would have some chance variability. Conditional on G and the two sample sizes, under the null hypothesis, the tickets in the first sample are like a random sample of size n1 without replacement from a collection of N = n1 + n2 units of which G are labeled with ones. Thus, under the null hypothesis, the number of tickets labeled with ones in the first sample has (conditional on G) an hypergeometric distribution with parameters N, G, and n1. Fisher's exact test uses this distribution to set the ranges of observed values of the number of ones in the first sample for which we would reject the null hypothesis.

■ **fixed-point free**
a flow or map is fixed-point free if it does not send any point to itself; any such point would be a fixed point.

■ **flip**
a transformation, also called a reflection, that produces a mirror image of a geometric figure.

■ **floor function**
the floor function of x is the greatest integer in x, i.e. the largest integer less than or equal to x.

■ **flow**
a vector field in Euclidean space or on a manifold which gives the velocity of a particle moving through the manifold as a function of its position. More technically, an autonomous ordinary differential equation.

■ **fluency**
the state of being able to smoothly and easily perform a given function or tasj.

■ **focal chord**
a chord of a conic that passes through a focus.

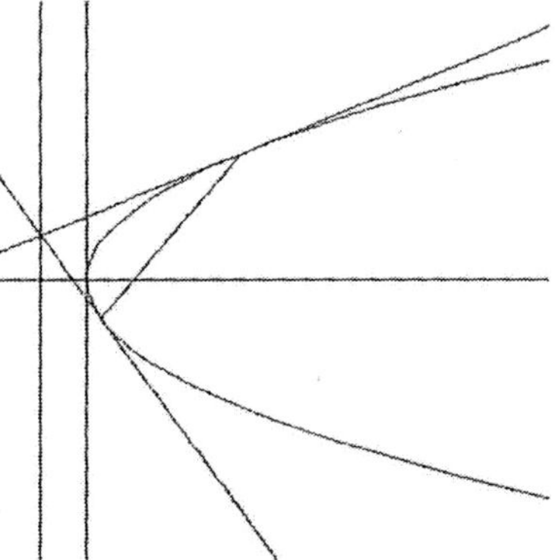

■ **focal radius**
a line segment from the focus of an ellipse to a point on the perimeter of the ellipse.

■ **focus**
an imaginary point used in determining the shapes of conic sections like parabolas, hyperbolas, and ellipses.

■ **foliation**
a decoration of a manifold in which the manifold is partitioned into sheets of some lower dimension, and the

sheets are locally parallel. (More technically, the foliated manifold is locally homeomorphic to a vector space decorated by cosets of a subspace).

foot of altitude

the intersection of an altitude of a triangle with the base to which it is drawn.

foot of line

the point of intersection of a line with a line or plane.

football-snaped scatterplot

in a football-shaped scatterplot, most of the points lie within a tilted oval, shaped more-or-less like a football. A football-shaped scatterplot is one in which the data are homoscedastically scattered about a straight line.

formula

1. a concise statement expressing the symbolic relationship between two or more quantities.

2. an equation or other mathematical expression used to calculate a value many times ($a^2 + b^2 = c^2$ is a formula).

Fourier series

a periodic function with period 2 pi.

fractal

an algebraically generated complex geometric shape having the property of being endlessly self-similar under magnification. Some computer screen savers utilise fractals.

fractals

geometrical entities characterised by basic patterns that are repeated at ever decreasing sizes. They are relevant to any system involving self-similarity repeated on diminished scales (such as a fern's structure) as in the study of chaos.

fraction

a way of representing part of a whole or part of a group by telling the number of equal parts in the whole and the number of those parts that are being described. For example, in the fraction 4/

5, there are 5 parts in the whole and 4 of those parts are being described.

■ **fraction paper strips**
strips of paper cut the same length and marked at various fraction locations.

■ **frame, sampling frame**
a sampling frame is a collection of units from which a sample will be drawn. Ideally, the frame is identical to the population we want to learn about; more typically, the frame is only a subset of the population of interest. The difference between the frame and the population can be a source of bias in sampling design, if the parameter of interest has a different value for the frame than it does for the population. For example, one might desire to estimate the current annual average income of 1998 graduates of the University of California at Berkeley. I propose to use the sample mean income of a sample of graduates drawn at random. To facilitate taking the sample and contacting the graduates to obtain income information from them, I might draw names at random from the list of 1998 graduates for whom the alumni association has an accurate current address. The population is the collection of 1998 graduates; the frame is those graduates who have current addresses on file with the alumni association. If there is a tendency for graduates with higher incomes to have up-to-date addresses on file with the alumni association, that would introduce a positive bias into the annual average income estimated from the sample by the sample mean.

■ **Fredholm determinant**
a complex analytic function which generalises the characteristic polynomial of a matrix. It is defined for those operators which have continuous kernels, i.e., kernels in the sense of analysis.

■ **frequency**
the number of times a value occurs in some time interval.

frequency table

a table listing the frequency (number) or relative frequency (fraction or percentage) of observations in different ranges, called class intervals.

frequency theory of probability

see Probability, Theories of.

front-ending

an estimating strategy that involves using the front end digits in a problem to arrive at an approximate value.

frustum

For a given solid figure, a related figure formed by two parallel planes meeting the given solid. In particular, for a cone or pyramid, a frustum is determined by the plane of the base and a plane parallel to the base. NOTE: this word is frequently incorrectly misspelled as frustrum.

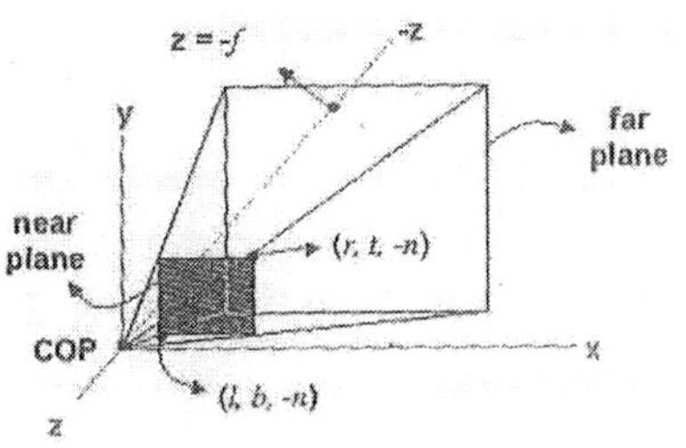

function (f)

the mathematical operation that transforms a piece of data into a different one. For example, f(x) = x2 is a function transforming any number to its square.

functional analysis

the study of differentiation, integration, estimates, and asymptotics of functions of real numbers. The modern research version of calculus.

fundamental group

a group, often non-abelian, that rather fully describes the periodicity or the 1-dimensional holes in a topological space.

fundamental rule of counting

if a sequence of experiments or trials T1, T2, T3, … , Tk could result, respectively, in n1, n2 n3, … , nk possible outcomes, and the numbers n1, n2 n3, … , nk do not depend on which outcomes actually occurred, the entire sequence of k experiments has n1× n2 × n3× . . . × nk possible outcomes.

■ G_2

the group of symmetries of the Cayley numbers. It is a compact, 14-dimensional Lie group which is not naturally part of an infinite sequence of Lie groups, hence it is an exceptional Lie group.

■ game theory

a field of study that bridges mathematics, statistics, economics, and psychology. It is used to study economic behavior, and to model conflict between nations, for example, "nuclear stalemate" during the Cold War.

■ gauge field/connection

a force field in nature, or an analogous vector field in mathematics with an enomormous amount of symmetry that expresses the redundancy or ambiguity of many parameters. The simplest example is the electromagnetic field; its inherent symmetry implies that in electrical circuits, relative voltages are measureable but absolute voltages are an arbitrary convention. The most interesting gauge fields have non-abelian symmetry groups. Also called a connection in differential geometry.

■ Gaussian curve

a normal curve.

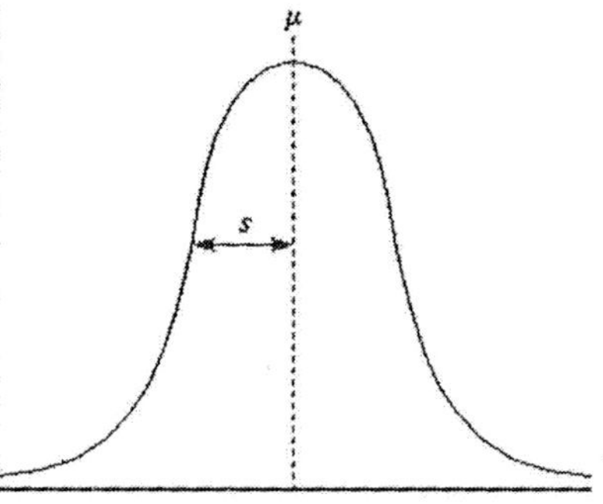

■ Gelfand-Fuchs cohomology

an ad hoc modification of certain homology {cohomology} groups in various infinite-dimensional settings that is more relevant and easier to compute than ordinary cohomology.

■ general relativity

einstein's model of gravity as the curvature of space and time. General relativity implies that spacetime is a 4-manifold.

general topology

the study of abstract, formal, and foundational properties of topological spaces. General topology is related to logic, functional analysis, and dynamical systems.

genus

1. in geometric topology, the number of holes of a surface. Usually this means the maximum number of disjoint circles that can be drawn on the surface such that the complement is connected. If there are no such circles, the surface is planar, and the genus then sometimes means the maximum number of disjoint arcs with the same property.
2. a measure of the twistedness of a fiber bundle. Technically, if X is a topological space with a fiber bundle F, it is the minimum number of open sets that cover X such that F is trivial over each open set.
3. Some other number which generalises or is analogous to the topological genus, such as the arithmetic genus of an algebraic curve.

geoboard

a flat board into which nails have been driven in a regular rectangular pattern. These nails represent the lattice points in the plane.

geodesic

1.on a Riemannian manifold or some other metric space, a curve which is the shortest path between any two points on it that are sufficiently close together.
2. the arc on a surface of shortest length joining two given points.

geodesy

a branch of mathematics dealing with the shape, size, and curvature of the Earth.

geometric distribution

the geometric distribution describes the number of trials up to and including the first success, in independent trials with the same probability of success. The geometric distribution depends only on the single parameter p, the probability of success in each trial. For example, the number of times one must toss a fair coin until the first time the coin lands heads has a geometric distribution with parameter p = 50%. The geometric distribution assigns probability p×(1 - p)k-1to the event that it takes k trials to the first success. The expected value of the geometric distribution is 1/p, and its SE is (1-p)½/p.

■ **geometric function theory**
the study of the geometry in the complex plane of complex analytic functions, in particular the relation between the image of the unit disk of an analytic function and its power series.

■ **geometric mean**
the geometric mean of two numbers is the square root of the product of the numbers. The geometric mean of n numbers is the nth root of the product of the numbers.

■ **geometric progression**
a sequence in which the ratio of each term to the preceding term is a given constant.

■ **geometric sequences**
a sequence in which there is a common ratio between successive terms. Each successive term of a geometric sequence is found by multiplying the preceding term by the common ratio. (e.g., in the sequence {1, 3, 9, 27, 81, ...} the common ratio is 3).

■ **geometric series**
the sum of a geometric sequence.

■ **geometric solid**
the bounding surface of a 3-dimensional portion of space.

■ **geometrical optics/ WKB approximation**
an approximation of a wave equation which assumes that the physical environment of a wave varies slowly compared to its wavelength. It predicts that the wave follows the trajectory of a refracting ray, such as a light ray.

■ **Geometrisation Con-**

jecture

the conjecture of Thurston that, after cutting along essential spheres and tori, every compact 3-manifold admits a special Riemannian metric known as a geometry, usually hyperbolic geometry. The conjecture subsumes the Poincaré conjecture and many other standard conjectures about 3-manifolds, and constitutes a classification of 3-manifolds.

■ **Geometrisation Theorem**

the Geometrisation Conjecture for Haken manifolds, proved by Thurston. (The technical heart of the theorem, the lemma that the skinning map has a unique fixed point, was independently proved by McMullen.).

■ **geometry**

the branch of mathematics that deals with the nature of space and the size, shape, and other properties of figures as well as the transformations that preserve these properties.

■ **Gergonne point**

in a triangle, the lines from the vertices to the points of contact of the opposite sides with the inscribed circle meet in a point called the Gergonne point.

■ **Glimm's method**

a method of proving existence of solutions to certain partial differential equations by discretely approximating space by a lattice and then taking the limit as the lattice spacing goes to zero.

■ **global analysis**

the study of partial differential equations (and other structures from analysis) on manifolds.

■ **gnomon magic square**

a 3 x 3 array in which the elements in each 2 x 2 corner have the same sum.

■ **Gödel, Kurt (1906 – 1978)**

the foremost logician of the 20th century. He is famous for many deep results in the foundations of mathematics, including the Godel Completeness Theorem, the Godel Incompleteness Theorem, and that the generalized continuum hypothesis and the axiom of choice are consistent with the other axioms of set theory. He moved from Austria to the Institute for Advanced Study at Princeton, New Jersey, in 1940. He was a lifelong friend of Albert Einstein, and in the 1950's formulated versions of Einstein's relativity theories that permitted the logical possibility of time travel.

Goldbach conjecture

every even number greater than 4 is the sum of two odd primes (32 = 13 + 19). Every odd number greater than 7 can be expressed as the sum of three odd prime numbers (11 = 3 + 3 + 5).

golden ratio

(1+Sqrt[5])/2.

golden rectangle

a rectangle whose sides are in the golden ratio.

graceful graph

a graph is said to be graceful if you can number the n vertices with the integers from 1 to n and then label each edge with the difference between the numbers at the vertices, in such a way that each edge receives a different label.

grad (or grade)

1/100th of a right angle.

gradient

the slope of a line. The gradient of two points on a line is calculated as rise (vertical increase) divided by run (horizontal increase).

Grand Unified Theory

a theory in fundamental physics that unifies the three forces of elementary particles: Electromagnetic forces, weak interactions, and strong interactions. The first are already unified in the electroweak theory. The electroweak and strong forces in ununified form, together with a short list of particles that interact via these forces, is called the Standard Model and may be a precursor to a Grand Unified Theory.

■ graph

a graph is a set of points (called vertices) and a set of lines (called edges) joinging these vertices.

■ graph of averages

for bivariate data, a graph of averages is a plot of the average values of one variable (say y) for small ranges of values of the other variable (say x), against the value of the second variable (x) at the midpoints of the ranges.

■ graph theory

the study of graphs, either for their own sake, or as models of such diverse things as groups (in pure mathematics) or computer networks.

■ great circle

a circle on the surface of a sphere whose centre is the centre of the sphere.

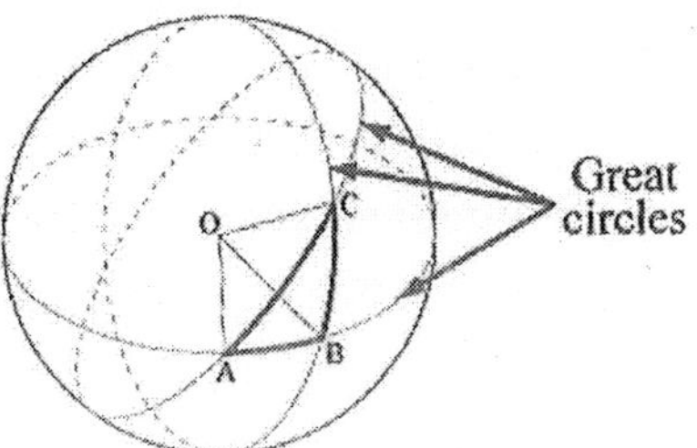

■ greatest common divisor

the greatest common divisor of a sequence of integers, is the largest integer that divides each of them exactly.

■ greatest common factor

same as greatest common divisor.

■ greatest common factor

the greatest common factor of two numbers, a and b, is the largest number that divides both a and b evenly.

■ greatest lower bound

the greatest lower bound of a set of real numbers, is the largest real number that is smaller than each of the numbers in the set.

■ **Gromov norm**

an invariant associated with the homology of a topological space that measures how many simplices are needed to represent a given homology class.

■ **group**

a mathematical system consisting of elements from a set G and a binary operation * such that.

x*y is a member of G whenever x and y are.

(x*y)*z=x*(y*z) for all x, y, and z, there is an identity element e such that e*x=x*e=e for all x, each member x in G has an inverse element y such that x*y=y*x=e.

■ **Haefliger structure**

a relatively complicated structure on an n-manifold which consists of a k-plane field plus auxiliary information derived from the diffeomorphism group of (n-k)-dimensional Euclidean space. A foliation has a parallel Haefliger structure just as it has a parallel plane field, but a Haefliger structure may have a singular locus where it is not parallel to a foliation.

■ **Haken**

a noted mathematician with the privilege of having an adjective named after him. A 3-manifold is Haken if it is closed, orientable, and has no essential sphere, but has an incompressible surface.

■ **half plane**

the set of all points in a plane that lie on one side of a line in the plane.

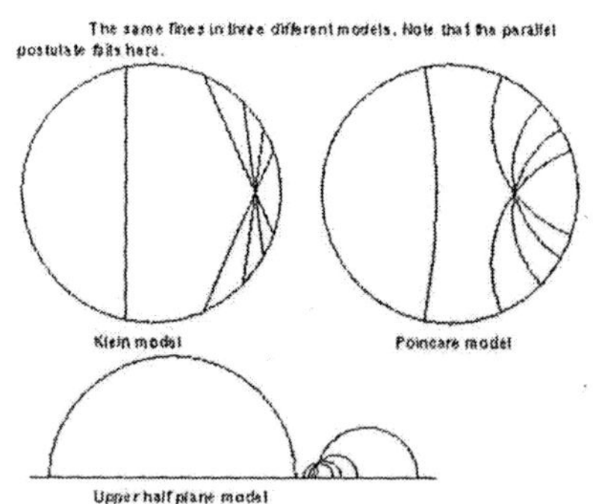

■ **half-line**

a ray.

■ **half-plane**

the part of a plane that lies on one side of a given line.

■ **Hamiltonian**

an expression that represents the equations of motion of a classical physical system via Hamilton's equations. The equations are related to the Hamiltonian via phase space or symplectic geometry. The solutions conserve the Hamiltonian if it is time-independent, and the value of the Hamiltonian can be interpreted as the energy of the system.

■ **Hankel matrix**

a matrix in which all the elements are the same along any diagonal that slopes from northeast to southwest.

$$H_i = \begin{bmatrix} m_i & m_{i+1} & \cdots & m_{i+(n-1)} \\ m_{i+1} & m_{i+2} & \cdots & \vdots \\ \vdots & \vdots & \ddots & \vdots \\ m_{i+(n-1)} & \cdots & \cdots & m_{i+2(n-1)} \end{bmatrix}$$

■ **harmonic analysis**

the study of the representation of functions by means of linear operations on characteristic sets of functions.

■ **harmonic division**

a line segment is divided harmonically by two points when it is divided externally and internally in the same ratio.

■ **harmonic mean**

of a set of numbers (y1 to yi), the harmonic mean is the reciprocal of the arithmetic mean of the reciprocal of the numbers [H = N / S (1/ y)].

■ **harmonic sequence**

a sequence is a harmonic sequence if the reciprocals of the numbers in the sequence form an arithmetic sequence.

■ **hectare**

a unit of measurement in the metric system equal to 10,000 square meters (approximately 2.47 acres).

■ **Heegaard splitting**

a division of a 3-manifold into two handlebodies. An early result in topology states that every closed 3-manifold (closed meaning that the manifold is finite and connected but has no boundary) has a Heegaard splitting and a resulting description in terms of a Heegaard diagram, which describes how the two handlebodies are glued together. The surface lying between the two handlebodies of the splitting is a Heegaard surface.

■ helix

the path followed by a point moving on the surface of a right circular cylinder that moves along the cylinder at a constant ratio as it moves around the cylinder. The parameteric equation for a helix is.

x=a cos t.

y=a sin t.

z=bt.

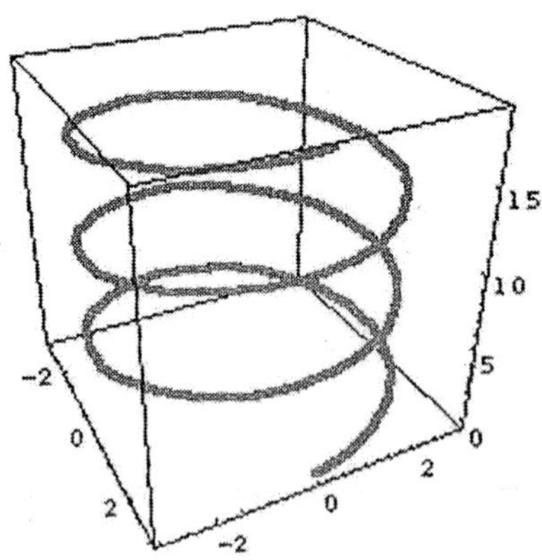

■ heptagon

a polygon with seven sides.

■ Heronian triangle

a triangle with integer sides and integer area.

■ Heron's formula

a formula for the area of a triangle.

a = sqrt[(s(s - a)(s - b)(s - c))]

where a, b, and c are the lengths of the sides of the triangle, and s is half the perimeter.

■ heteroscedasticity

"Mixed scatter." A scatterplot or residual plot shows heteroscedasticity if the scatter in vertical slices through the plot depends on where you take the slice. Linear regression is not usually a good idea if the data are heteroscedastic.

■ heuristic

rules, suggestions, guides, or techniques that may be useful in making progress toward a solution of a problem.

■ hexadecimal number

a number written in base sixteen.

■ **hexagon**
a polygon with six sides.

■ **hexagonal number**
a number of the form n(2n-1).

■ **hexagonal prism**
a prism with a hexagonal base.

■ **hexahedron**
a polyhedron with six faces. A regular hexahedron is a cube.

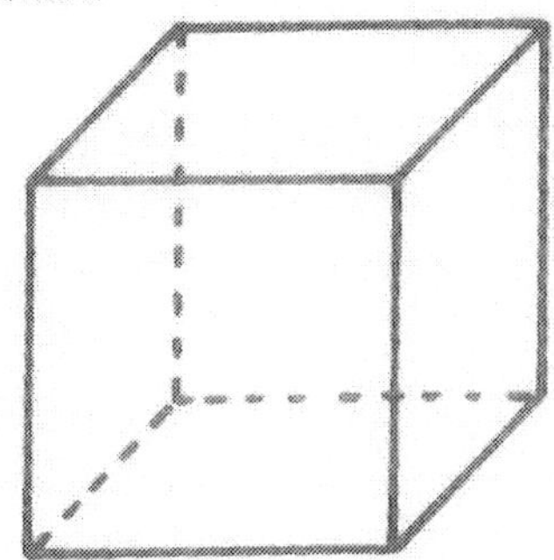

■ **hexomino**
a six-square polyomino.

■ **hierarchy**
an infinite system of equations whose truncations (meaning the first n equations for some n) are meaningful and have similar properties. Usually hierarchies refer to systems of partial differential equations generalising a single equation of interest. For example, the KP hierarchy generalises the KP equation.

■ **hierarchy of operations**
in an equation with multiple operators, operations proceed in the following order exponentiation, division/multiplication, subtraction/summation and from left to right.

■ **highest common factor (HCF)**
the greatest natural number which is a factor of two or more given numbers.

■ **histogram**
a histogram is a kind of plot that summarises how data are distributed. Starting with a set of class intervals, the histogram is a set of rectangles ("bins") sitting on the horizontal axis. The bases of the rectangles are the class intervals, and their heights are such that their areas are proportional to the fraction of observations in the corresponding class intervals. That is, the height of a

given rectangle is the fraction of observations in the corresponding class interval, divided by the length of the corresponding class interval. A histogram does not need a vertical scale, because the total area of the histogram must equal 100%. The units of the vertical axis are percent per unit of the horizontal axis. This is called the density scale. The horizontal axis of a histogram needs a scale. If any observations coincide with the endpoints of class intervals, the endpoint convention is important.

■ historical controls

sometimes, the treatment group is compared with individuals from another epoch who did not receive the treatment; for example, in studying the possible effect of fluoridated water on childhood cancer, we might compare cancer rates in a community before and after fluorine was added to the water supply. Those individuals who were children before fluoridation started would comprise an historical control group. Experiments and studies with historical controls tend to be more susceptible to confounding than those with contemporary controls, because many factors that might affect the outcome other than the treatment tend to change over time as well. (In this example, the level of other potential carcinogens in the environment also could have changed.).

■ homeomorphism

1.a one-to-one continuous transformation that preserves open and closed sets. 2.a function that preserve the operators associated with the specified structure.

■ homological algebra

the algebraic study of the homology and cohomology of manifolds and other mathematical objects. Homological algebra is a grand generalisation of linear algebra.

■ homology/cohomology

homology and cohomology are algebraic objects associated to a manifold or other mathematical object which give one measure of the number of holes of the object. The homology of a topological space has a relatively technical definition, but it is relatively easy to compute and study with tools from linear algebra.

homoscedasticity

"Same scatter." A scatterplot or residual plot shows homoscedasticity if the scatter in vertical slices through the plot does not depend much on where you take the slice. C.f. heteroscedasticity.

Hopf algebra

an abstract algebraic object, generalising a group or a Lie algebra, with enough structure to have a representation theory.

horocycle/horosphere

a generalised (round) circle or sphere in the hyperbolic plane or space. A horosphere has infinite radius and meets the sphere at infinity at one point, and its geometric centre is the same point.

house edge

in casino games, the expected payoff to the bettor is negative the house (casino) tends to win money in the long run. The amount of money the house would expect to win for each $1 wagered on a particular bet (such as a bet on "red" in roulette) is called the house edge for that bet.

HTLWS

the book How to lie with Statistics by D. Huff.

hyperbola

1. a curve with equation

$$\frac{x^2}{a^2} - \frac{y^2}{b^2} = 1.$$

2. a curved line where the difference of the distances from two points (foci) to each point on the curve is constant.

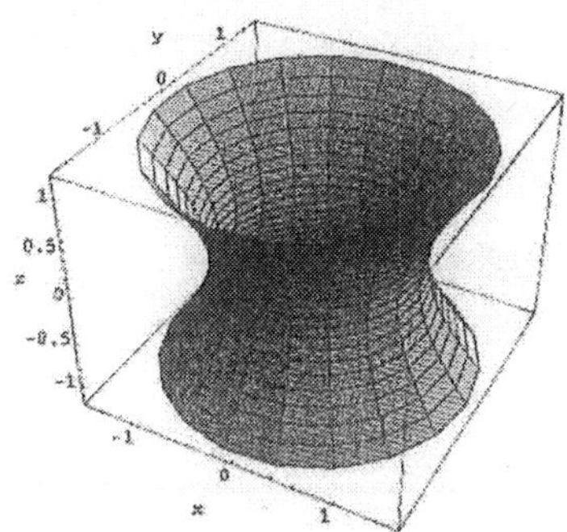

hyperbolic cosine

$\cosh x = (1/2)(e^{x} + e^{-x})$.

hyperbolic functions

the hyperbolic functions are defined as follows.

hyperbolic geometry

in modern terms, hyperbolic geometry is the study of manifolds with Riemannian metrics with constant negative curvature. The hyperbolic plane is a particular hyperbolic manifold which is in a sense universal among hyperbolic surfaces; similarly there is also hyperbolic n-space. Escher's circle limit prints are excellent illustrations of the hyperbolic plane.

hyperbolic PDE

a partial differential equation that resembles a wave equation with one time coordinate, one or several space coordinates, and a finite speed of propogation for features of solutions.

hyperbolic sine

$\sinh x = (1/2)(e^{x} - e^{-x})$.

hyperbolic spiral

the curve whose equation in polar coordinates is $r\theta = a$.

hyperbolic tangent

tanh x = sinh x.

hyperboloid

a geometric solid whose equation is

$$\frac{x^2}{a^2} + \frac{y^2}{b^2} - \frac{z^2}{c^2} = 1$$

hypergeometric distribution

the hypergeometric distribution with parameters N, G and n is the distribution of the number of "good" objects in a simple random sample of size n (i.e., a random sample without replacement in which every subset of size n has the same chance of occurring) from a population of N objects of which G are "good." The chance of getting exactly g good objects in such a sample is.

gCg × N-GCn-g/NCn,

provided g <= n, g <= G, and n - g <= N - G. (The probability is zero otherwise.) The expected value of

the hypergeometric distribution is n×G/N, and its standard error is.
((N-n)/(N-1))½ × (n × G/ N × (1-G/N))½.

hyperplane/hypersurface

a high-dimensional plane or submanifold in a vector space or manifold with codimension 1.

hypotenuse

the longest side of a right triangle which lies opposite the vertex of the right angle.

hypothesis testing

statistical hypothesis testing is formalized as making a decision between rejecting or not rejecting a null hypothesis, on the basis of a set of observations. Two types of errors can result from any decision rule (test) rejecting the null hypothesis when it is true (a Type I error), and failing to reject the null hypothesis when it is false (a Type II error). For any hypothesis, it is possible to develop many different decision rules (tests). Typically, one specifies ahead of time the chance of a Type I error one is willing to allow. That chance is called the significance level of the test or decision rule. For a given significance level, one way of deciding which decision rule is best is to pick the one that has the smallest chance of a Type II error when a given alternative hypothesis is true. The chance of correctly rejecting the null hypothesis when a given alternative hypothesis is true is called the power of the test against that alternative.

i

the square root of -1.

icosahedron

a polyhedron with 20 faces.

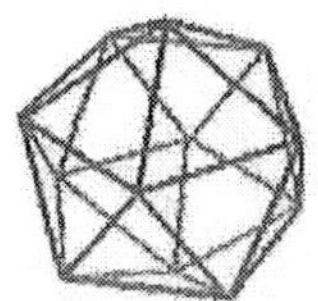

20 faces
30 edges
12 vertices

regular icosahedron

icosohedron

a polyhedron with 20 faces.

ideal vertex

a vertex at infinity of a hyperbolic polyhedron. Geometrically, the faces of the polyhedron taper together at an exponential rate as they extend towards the ideal vertex.

idempotent

the element x in some algebraic structure is called idempotent if x*x=x.

identity

an equation that is true for all values of the variable.

identity element

the element of a set which when combined with any element of the same set leaves the other element unchanged (like zero in addition and subtraction, and 1 in multiplication or division).

identity matrix

a square matrix with ones along the diagonal and zeros everywhere else. If I is an identity matrix, then IA = A.

$$I_4 = \begin{bmatrix} 1 & 0 & 0 & 0 \\ 0 & 1 & 0 & 0 \\ 0 & 0 & 1 & 0 \\ 0 & 0 & 0 & 1 \end{bmatrix}$$

iff, if and only if

if p and q are two logical propositions, then(p IFF q) is a proposition that is true when both p and q are true, and when both p and q are false. If is logically equivalent to the proposition.

((p IMPLIES q) AND (q IMPLIES p)) and to the proposition ((p AND q) OR ((NOT p) AND (NOT q))).

imaginary axis

the y-axis of an Argand diagram.

imaginary number

1.a complex number of the form xi where x is real and i=sqrt(-1).

2.any positive root of a negative number, as in the fourth root of -3.

imaginary number

the product of a real number x and i, where i2 + 1 = 0. A complex number in which the real part is zero. In general, imaginary numbers are the square roots of negative numbers.

imaginary part

the imaginary part of a complex number x+iy where x and y are real is y.

implication

a conditional statement.

implies, logical implication

Logical implication is an operation on two logical propositions. If p and q are two logical propositions, (p IMPLIES q) is a logical proposition that is true if p is false, or if both p and q are true. The proposition (p IMPLIES q) is logically equivalent to the proposition ((NOT p) OR q).

improper fraction

A fraction whose numerator is the same as or larger than the denominator; i.e., a fraction equal to or greater than 1.

incenter

the center of a circle that is inscribed in a triangle. The intersection of the angle bisectors of the triangle.

incircle

the circle that can be inscribed in a triangle.

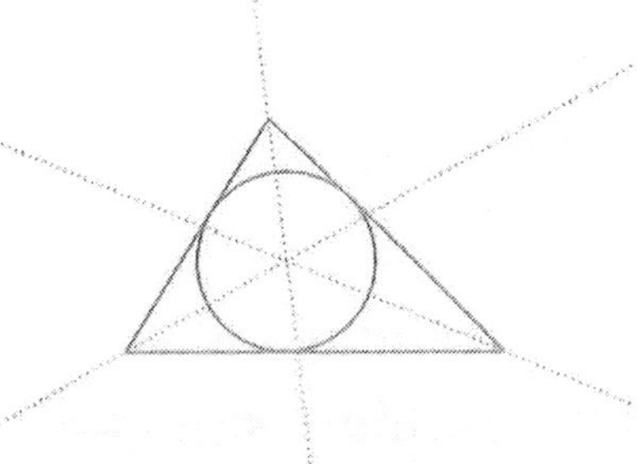

incompressible

1. in physics, refers to a fluid that cannot change volume or a flow that is possible for such a fluid.

2. in 3-dimensional topology, refers to a surface in a 3-manifold with the property that no essential circle in the surface bounds a disk in the manifold.

incompressible fluid

a fluid that does not change its volume except under extreme conditions, or a mathematically abstracted fluid that strictly conserves its volume.

increasing function

a function is increasing if f(a) > f(b) when a > b.

increment

a small change, usually indicated by the greek letter delta.

indefinite integral

the sum of the antiderivative of a function and an arbitrary constant.

independent and identically distributed (iid)

a collection of two or more random variables {X1, X2, ... , } is independent and identically distributed if the variables have the same probability distribution, and are independent.

independent variable

in regression, the independent variable is the one that is supposed to explain the other; the term is a synonym for "explanatory variable." Usually, one regresses the "dependent variable" on the "independent variable." There is not always a clear choice of the independent variable. The independent variable is usually plotted on the horizontal axis. Independent in this context does not mean the same thing as statistically independent.

independent, independence

two events A and B are (statistically) independent if the chance that they both happen simultaneously is the product of the chances that each occurs individually; i.e., if P(AB) = P(A)P(B). This is essentially equivalent to saying that learning that one event occurs does not give any information about whether the other event occurred too the conditional probability of A given B is the same as the unconditional probability of A, i.e., P(A|B) = P(A). Two random variables X and Y are independent if all events they determine are independent, for example, if the event {a < X <= b} is independent of the event {c < Y

<= d} for all choices of a, b, c, and d. A collection of more than two random variables is independent if for every proper subset of the variables, every event determined by that subset of the variables is independent of every event determined by the variables in the complement of the subset. For example, the three random variables X, Y, and Z are independent if every event determined by X is independent of every event determined by Y and every event determined by X is independent of every event determined by Y and Z and every event determined by Y is independent of every event determined by X and Z and every event determined by Z is independent of every event determined by X and Y.

■ indicator random variable

the indicator [random variable] of the event A, often written 1A, is the random variable that equals unity if A occurs, and zero if A does not occur. The expected value of the indicator of A is the probability of A, P(A), and the standard error of the indicator of A is $(P(A)\times(1-P(A))^{1/2}$. The sum.

1A + 1B + 1C + …

of the indicators of a collection of events {A, B, C, … } counts how many of the events {A, B, C, … } occur in a given trial. The product of the indicators of a collection of events is the indicator of the intersection of the events (the product equals one if and only if all of indicators equal one). The maximum of the indicators of a collection of events is the indicator of the union of the events (the maximum equals one if any of the indicators equals one).

■ indirect proof

a deductive proof using contradiction or elimination to rule out all except the desired conclusion.

■ inductive reasoning

a form of reasoning from individual cases to general

ones or from observed instances to unobserved ones.

■ **inequality**

1.the statement that one quantity is less than (or greater than) another.

2.an expression with two unequal values separated by a less-than or greater-than sign, like 4

■ **infinite**

1.becoming large beyond bound.

2. a series or number which has no end and continues forever. infinitesimal A vanishingly small part of a quantity. It equals almost zero.

■ **infinity**

a limitless quantity.

■ **inflection**

in mathematics, the concept of an infinite-dimensional space considered literally. It is a vector space with an infinite basis or a space with infinitely many coordinates.

■ **inflection point**

a variable that approaches 0 as a limit.

■ **initial condition**

the state of a system at its initial time of consideration. Equivalently, a prespecified value or set of values of a function satisfying a differential equation at an initial time.

■ **injection**

a one-to-one mapping.

■ **inscribed angle**

the angle formed by two chords of a curve that meet at the same point on the curve.

■ **inscribed polygon**

a polygon placed inside a circle so that each vertex of the polygon touches the circle.

■ **instanton**

a solution to a partial differential equation, especially a non-abelian gauge field equation, which is localised in all directions.

■ **integer**

1.one of the numbers ..., -3, -2, -1, 0, 1, 2, 3, ...

2.a number in the set of positive and negative whole numbers as well as zero. {...,-2,-1,0,1,2,...}

integrable system
an ordinary or partial differential equation or system of equations with a maximal number of conserved quantities such as energy and momentum. In a sense, an integrable system can be completely solved.

integral
if $dF(x)/dx = f(x)$, then $F(x)$ is an integral of $f(x)$. The area under the curve of a function above the x - axis.

integral calculus
this is the inverse process to differentiation. For example, x^2 has derivative $2x$, so $2x$ has x^2 as an integral. A classic application of integral is to calculate areas.

integrand
a function that is to be integrated.

integrate/integrable
an infinitesimal structure of any of various kinds can often be integrated, meaning pieced together into a geometric whole. This process is usually abstracted from the prime example: integrating a function in calculus. Typically the infinitesimal structure could have pervasive inconsistencies that would prevent it from being integrable.

integration
the process of finding an integral.

integro-differential equation
an equation relating a function to both its derivatives and antiderivatives.

interacting particle system
a dynamical system consisting of a large number of particles which collide or otherwise interact when they are close to each other. For example, a microscopic model of a molecular liquid or gas.

intercept
the x-intercept of a curve is the point where the curve crosses the x - axis, and the y - intercept of a curve is the point where the curve crosses the y - axis.

interpolate

in a table of numerical values, any procedure for estimating intermediate values that are not listed. When estimating from a graph, the estimation of a value between given points.

interpolation

given a set of bivariate data (x, y), to impute a value of y corresponding to some value of x at which there is no measurement of y is called interpolation, if the value of x is within the range of the measured values of x. If the value of x is outside the range of measured values, imputing a corresponding value of y is called extrapolation.

inter-quartile range (iqr)

the inter-quartile range of a list of numbers is the upper quartile minus the lower quartile.

intersection

the intersection of two or more sets is the set of elements that all the sets have in common; the elements contained in every one of the sets. The intersection of the events A and B is written "A and B" and "AB." C.f. union. See also Venn diagrams.

invariant

in topology, a number, polynomial, or other quantity associated to a topological object such as a knot or 3-manifold which depends only on the underlying object and not on its specific description or presentation.

inverse

a related but opposite process or number such as multiplication being the inverse of division and 2/1 being the inverse of 1/2.

inversely proportional

y is inversely proportional to x if $y = k/x$.

irrational number

a real number that cannot be expressed as the ratio of two integers, and therefore that cannot be written as a decimal that either terminates or repeats. The square root of 2 is an example because if it is expressed as a ratio, it never gives 2 when multiplied by itself. The numbers p = 3.141592645..., and e = 2.7182818... are also irrational numbers.

ising model

the first and simplest interesting model in statistical mechanics. It consists of a grid of (abstracted) atoms, each of which can be in either of two states, spin up or spin down. The energy of a given state of the grid is given by the number of atoms which are spin up or down and by the number of pairs of neighbouring atoms whose spins agree or disagree.

isogonal conjugate

isogonal lines of a triangle are cevians that are symmetric with respect to the angle bisector. Two points are isogonal conjugates if the corresponding lines to the vertices are isogonal.

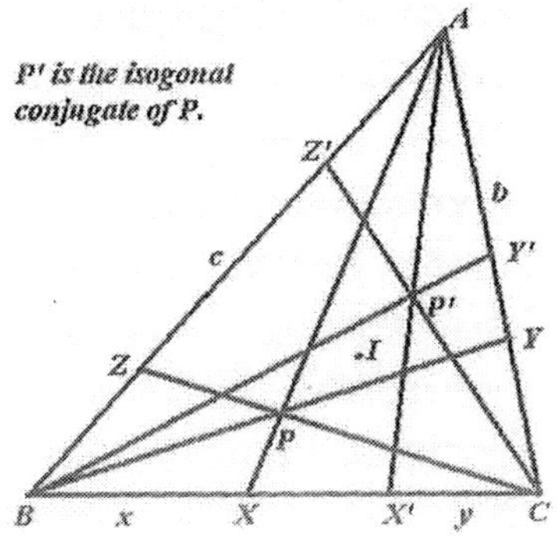

isometry

a transformation of a figure that does not change the distances of any two points in the figure.

isosceles tetrahedron

a tetrahedron in which each pair of opposite sides have the same length.

isosceles trapezoid

ain which the two non-parallel sides have the same length.

isospectral deformation

a perturbation (continuous change) of the entries of a matrix, or the coefficients of a differential operator or other linear operator, that does not change its eigenvalues.

isotomic conjugate
two points on the side of a triangle are isotomic if they are equidistant from the midpoint of that side. Two points inside a triangle are isotomic conjugates if the corresponding cevians through these points meet the opposite sides in isotomic points.

iteration
repeatedly performing the same sequence of steps. Simply, solving an algebraic equation with an arbitrary value for the unknown and using the result to solve it again, and again.

iterative processes
in discrete math, a method of calculating an amount by using an initial value and applying a function repeatedly.

Jacobian variety
a space associated to a Riemann surface defined most succinctly as the complex cohomology (as a vector space) divided by the integral cohomology (as a lattice). It is simultaneously a complex manifold, an algebraic variety, and a Lie group.

joint probability distribution
if X1, X2, … , Xk are random variables, their joint probability distribution gives the probability of events determined by the collection of random variables for any collection of sets of numbers {A1, … , Ak}, the joint probability distribution determines.

p((X1 is in A1) and (X2 is in A2) and … and (Xk is in Ak)).

joint probability function
a function that gives the probability that each of two or more random variables takes at a particular value.

joint variation
a variation in which the values of one variable depend upon those of 2 or more variables.

■ joint variation

y varies jointly as x and z if y = kxz.

■ Jones polynomial

a famous invariant of knots and links discovered by Vaughan Jones. It has several extremely elementary definitions and at the same time involves deep mathematics.

■ Jordan curve

a simple closed curve.

Insideness- Jordan Curve Theorem
Shoot a ray from the point andcount the number of intersections with the boundary.
Odd - inside; *Even* - outside;

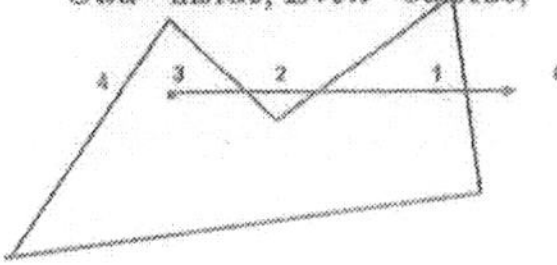

■ Jordan matrix

a matrix whose diagonal elements are all equal (and non-zero) and whose elements above the principal diagonal are equal to 1, but all other elements are 0.

■ joule

a unit of energy or work.

■ jump discontinuity

a discontinuity in a function where the left and righ-hand limits exist but are not equal to each other.

■ Kepler, Johannes (born 1571)

german astronomer and mathematician who discovered that planetary motion is elliptical. Early in his life, Kepler set about to prove that the universe obeyed Platonistic mathematical relationships; that for instance the planetary orbits were circular and at distances from the sun proportional to the Platonic solids. However, when his friend the astronomer Tycho Brahe died he bequeathed to Kepler his immense collection of astronomical observations. After decades of studying these observations, Kepler reaiized that his earlier surmises about planetary motion were naive, and he formulated his three laws of planetary motion, now known as Kepler's Laws. He did not have a uni fying theory for these laws, however; this had to wait another generation, until Isaac Newton formulated his laws of gravity and motion.

■ **kernel**

1. in algebra, the set of vectors annihilated (sent to zero) by a matrix, linear operator, homomorphism, or any similar function.

2. in analysis, a continuous analogue of a matrix. Given a vector space of functions of a parameter or functions on a manifold, an operator may have a kernel or matrix whose rows and columns are indexed by the parameter or by points on the manifold.

■ **kilo**

a prefix meaning one thousand.

■ **kilometer**

a unit of length equal to 1,000 meters.

■ **kinematics**

a branch of mechanics dealing with the motion of rigid bodies without reference to their masses or the forces acting on the bodies.

■ **kite**

a quadrilateral which has two pairs of adjacent sides equal.

■ **Klein-Gordon equation**

the partial differential equation Nabla u + u = 0 in Euclidean space or Minkowski space or a variant. The Klein-Gordon equation is the second-simplest PDE which is invariant under isometries of space; it arises in classical and quantum field theory in physics.

■ **knight's tour**

a knight's tour of a chessboard is a sequence of moves by a knight such that each square of the board is visited exactly once.

■ **knot**

1.a curve in space formed by interlacing a piece of string

and then joining the ends together.
2. a unit of speed in navigation equal to one nautical mile per hour.

Korteweg-de Vries equation/KdV equation

the partial differential equation $u_t + uu_x + u_{xxx} = 0$ which is important both in applied mathematics, because of the physical phenomena it models, and in pure mathematics, because of the structure of its solutions.

KP equation

the partial differential equation $u_{yy} - (u_t - u_{xxx} - uu_x)_x = 0$, an important integrable system in mathematical physics.

hospital's rule

if a limit is in one of the indeterminate forms "infinity over infinity" or "zero over zero," then we have.
that is, the limit is the same after taking the derivatives of both the numerator and denominator. L'Hospital's Rule may be applied as many times as needed. Students often misapply the Rule by using it when the limit is not in one of the above indeterminate forms. This often results in an incorrect evaluation of the limit.

Lagrangian

an expression that represents the equations of motion of a classical physical system via the Euler-Lagrange equations. The solutions to the Euler-Lagrange equations and the dynamics of the system correspond to minima, maxima, and other critical points of the Lagrangian.

lake and island board

a board (bristol) with appropriate geometric shapes attached. Used for measurement of perimeter and area.

lamination

a decoration of a manifold in which some subset is partitioned into sheets of some lower dimension, and the sheets are locally parallel. It may or may not be possible to fill the gaps in a lamination to make a foliation.

large deviations/ theory of large deviations

a collection of methods for estimating and proving results about astronomically unlikely events in probability theory, usually large deviations from likely outcomes dictated by the law of large numbers and the central limit theorem.

latera recta

plural of lattice rectum.

latin square

an n X n array of numbers in which only n numbers appear. No number appears more than once in any row or column.

latitude

the angular distance of a point on the Earth from the equator, measured along the meridian through that point.

lattice

a periodic arrangement of points such as the vertices of a tiling of space by cubes or the positions of atoms in a crystal. More technically, a discrete abelian subgroup of an n-dimensional vector space which not contained in an n-1-dimensional vector space. Lattices play a central role in the theory of Lie groups, in number theory, in error-correcting codes, and many other areas of mathematics.

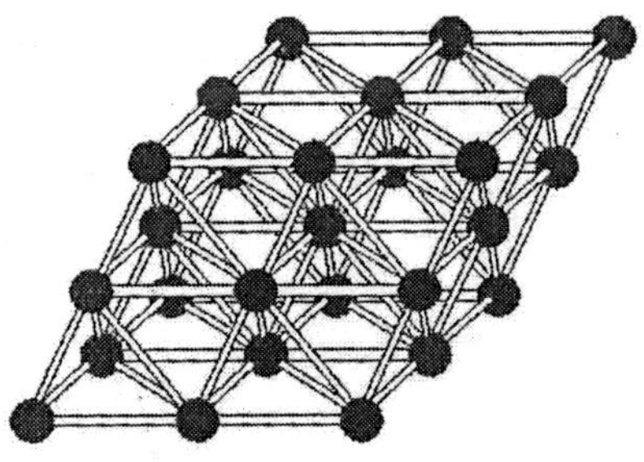

lattice path

a sequence of points in a lattice such that each point differs from its predecessor by a finite list of allowed steps. Random lattice paths are an interesting model for the random motion of a particle and lattice paths are also important in enumerative combinatorics.

lattice point

a point with integer coordinates.

latus rectum

a chord of an ellipse passing through a focus and perpendicular to the major axis of the ellipse.Plural: latera recta.

law of averages

the Law of Averages says that the average of independent observations of random variables that have the same probability distribution is increasingly likely to be close to the expected value of the random variables as the number of observations grows. More precisely, if X1, X2, X3, . . . , are independent random variables with the same probability distribution, and E(X) is their common expected value, then for every number E > 0,

p{|(X1 + X2 + … + Xn)/n - E(X) | < E}

converges to 100% as n grows. This is equivalent to saying that the sequence of sample means.

x1, (X1+X2)/2, (X1+X2+X3)/3, … converges in probability to E(X).

law of cosines

$c^2 = a^2 + b^2 - 2ab \cos c$

law of large numbers

the Law of Large Numbers says that in repeated, independent trials with the same probability p of success in each trial, the percentage of successes is increasingly likely to be close to the chance of success as the number of trials increases. More precisely, the chance that the percentage of successes differs from the probability p by more than a fixed positive amount, E > 0, converges to zero as the number of trials n goes to infinity, for every number e > 0. Note that in contrast to the difference between the percentage of successes and the probability of success, the difference between the number of successes and the expected number of successes, n×p, tends to grow as n grows. The following tool illustrates the law of large numbers; the button toggles between displaying the difference between the number of successes and the expected number of successes, and the difference between the percentage of successes and the expected percentage of successes. The tool on this page illustrates the law of large numbers.

law of sines

a/sin A = b/sin B = c/sin C.

least common denominator

the least common denominator of two fractions, a/b and c/d, is the smallest number that contains both b and d as factors.

least common multiple

the least common multiple of two numbers, a and b, is the smallest number that contains both a and b as factors.

least upper bound

the least upper bound of a set of numbers is the smallest number that is larger than every member of the set.

least-area surface

a surface (or manifold) lying in space or in a manifold which minimises area (surface volume) among a class of similar surfaces. Example: The round sphere has least area among surfaces in Euclidean space that enclose a fixed volume.

Leibniz, Gottfried Wilhelm (1646-1716)

german mathematician and philosopher, and co-discoverer of calculus (independently of Newton). Leibniz sought to reduce the mechanics of human thought to a logical calculus, and in this respect foreshadowed modern interest in artificial intelligence and cognitive science. He also described the binary number system and laid the foundation of dynamics. Leibniz's philosophy was characterized by the assertion that all true propositions are analytic, but that some propositions (which seem synthetic to us) could only be known to be analytic by God.

lemata

plural of lemma.

lemma

a proposition that is useful mainly for the proof of some other theorem.

leopard spot

in the proof of the Geometrisation Theorem, when the skinning map is applied to a conformal structure A to produce a new one B, B is made up of many (unrolled) copies of A that resemble leopard spots.

level spacing

in mathematical physics, the difference between consecutive elements in some set of real numbers, in particular the difference between consecutive energy levels or eigen values of a matrix or linear operator.

Lie algebra

an algebraic structure on a vector space which describes multiplication of elements of a Lie group which are very close to the identity (infinitesimal transformations). Lie algebras are almost as important as their comrades-at-arms, Lie groups.

Lie group

a group (in the sense of abstract algebra) which is at the same time a manifold. Example: The group of rotations in n dimensions is a Lie group of dimension n(n-1)/2. Lie groups are fundamental objects in mathematics and physics, especially quantum field theory.

life/Conway's game of Life

a popular 2-dimensional cellular automaton in which cells may be present or absent, a cell is born if in the previous iteration three cells (out of the nearest eight neighbours) were adjacent to an empty space, and a cell stays alive if two or three living cells were adjacent.

■ **like terms**
two terms each of whose parts, with the exception of their coefficients, is the same.

■ **limit**
see converge.

■ **line**
a straight set of points that extends off into infinity in both directions.

■ **line graph**
a diagram using line segments to represent a set of data.

■ **line segment**
the part of a line between two given distinct points on that line (including the two points).

■ **line symmetry**
when an object matches itself when reflected in a line.

■ **linear**
a model or function where the input and output are proportional.

■ **linear association**
two variables are linearly associated if a change in one is associated with a proportional change in the other, with the same constant of proportionality throughout the range of measurement. The correlation coefficient measures the degree of linear association on a scale of -1 to 1.

■ **linear differential operator**
an operator which takes functions to functions by taking a linear combination of different derivatives. For example, d/dt + t d2/dt2. It is an ordinary differential operator if there is only a single independent variable, i.e., if it corresponds to ordinary differential equations.

■ **linear equation**
an equation containing linear expressions.

linear expression

an expression of the form $ax+b$ where x is variable and a and b are constants; or in more variables, an expression of the form $ax + by + c$, $ax + by + cz + d$, etc.

linear function

a function that has a constant rate of change and can be modeled by a straight line.

linear operation

suppose f is a function or operation that acts on things we shall denote generically by the lower-case Roman letters x and y. Suppose it makes sense to multiply x and y by numbers (which we denote by a), and that it makes sense to add things like x and y together. We say that f is linear if for every number a and every value of x and y for which f(x) and f(y) are defined, (i) $f(a\times x)$ is defined and equals $a\times f(x)$, and (ii) $f(x + y)$ is defined and equals $f(x) + f(y)$. C.f. affine.

linear programming

the problem, and associated area of mathematics, of maximising or minimising a linear function on a convex set, especially a polytope. Equivalently, maximising a linear expression in some number of variables subject to linear equalities and inequalities.

link

a link is a collection of disjoint circles lying in a 3-manifold, often but not always Euclidean 3-space or the 3-sphere. A knot is link which happens to be a single circle. Manifold topologists usually study tame knots and links, which can be represented by smooth or polygonal curves, but there are also wild links which are infinitely knotted. Ordinary knots and links were topologically classified (in a certain sense) by Haken and Thurston.

literal numbers

letters representing numbers (as in algebraic equations).

localised solution
a solution of a differential equation, or a similar mathematical object, which is confined to a small region even though it has the freedom to spread out, like a lump in a carpet.

locally convex
for a topological vector space, the property that there exist arbitrarily small open regions around the origin (and consequently any other point) which are convex sets.

location, measure of
a measure of location is a way of summarizing what a "typical" element of a list is—it is a one-number summary of a distribution. See also arithmetic mean, median, and mode.

locus
the set of all points meeting some specified condition.

logarithm
the logarithm of a number N to a given base b is the power to which the base must be raised to produce the number N. Written as logb N. Naturally, logb bx = x. In any base, the following rules apply log (ab) = log a + log b; log (a/b) = log a - log b; log (1/a) = -log a; log ab = b log a; log 1 = 0 and log 0 is undefined.

logic
the branch of mathematics in which mathematical assertions and reasoning are studied as formal mathematical objects.

longitudinal study
a study in which individuals are followed over time, and compared with themselves at different times, to determine, for example, the effect of aging on some measured variable. Longitudinal studies provide much more persuasive evidence about the effect of aging than do cross-sectional studies.

long-time solution
a solution to a differential equation that exists, or has been proven to exist, for a long interval of the equation's time parameter, although not necessarily for infinite time.

■ loop and sphere theorems

two foundational theorems in 3-manifold topology, proved by Papakyriakopoulos in the 1950's. The loop and sphere theorems generalise Dehn's lemma, which asserts that if a knot in three dimensions bounds a disk which intersects itself in its interior, the disk can be simplified to remove all self-intersection and the knot must be trivial.

■ loop space

given a topological space X, its loop space is the topological space of all continuous functions from a circle to X. Loop spaces are important examples of new topological spaces formed from old ones, as well as examples of infinite-dimensional spaces in mathematics.

■ lower quartile (lq)

see quartiles.

■ lowest common denominator

the smallest number that is exactly divisible by each denominator of a set of fractions.

■ lowest common multiple (LCM).

the smallest non-zero natural number that is a common multiple of two or more natural numbers (compare with the highest common factor).

■ loxodrome

on a sphere, a curve that cuts all parallels under the same angle.

■ lozenge

a rhombus with a 60 degree angle.

■ L-shaped method

a method for two-stage linear stochastic programming problems in which one solves for all optimal recourses in bunches and uses the solutions to form a linear programming problem for the initial choice. It is based on the Benders decomposition method for mixed linear programming problems.

L-tetromino

a tetromino in the shape of the letter L.

Lucas number

a member of the sequence 2, 1, 3, 4, 7,... where each number is the sum of the previous two numbers. L0=2, L1=1, Ln=Ln-1+Ln-2.

lune

the portion of a sphere between two great semicircles having common endpoints (including the semicircles).

MacDonald identities

a set of simple formulas conjectured by MacDonald in the 1970's which give a signed enumeration of lattice paths in certain lattices related to Lie algebras. It was understood from computer evidence that they were almost certainly true, but a conceptual proof of the most general identities had to await sophisticated ideas in the theory of infinite-dimensional Lie algebras.

Mach reflection

in ordinary reflection of waves, the incoming and reflected waves fronts make a V shape. Mach reflection is a non-linear type of reflection where the two waves merge a distance away from the surface of reflection to make a Y.

magic square

a square array of n numbers such that sum of the n numbers in any row, column, or main diagonal is a constant (known as the magic sum).

23	6	19	2	15
10	18	1	14	22
17	5	13	21	9
4	12	25	8	16
11	24	7	20	3

magic tour

if a chess piece visits each square of a chessboard in

succession, this is called a tour of the chessboard. If the successive squares of a tour on an n X n chessboard are numbered from 1 to n2, in order, the tour is called a magic tour if the resulting square is a magic square.

magnitude
the magnitude of a vector is its length.

main diagonal
in the matrix [aij], the elements a11, a22, ..., ann.

major arc
an arc of measure greater than 180 degrees.

major axis
the major axis of an ellipse is it's longest chord.

Malfatti circles
three equal circles that are mutually tangent and each tangent to two sides of a given triangle.

Mandelbrot set
a certain subset of the complex plane which is ubiquitous in popular accounts of research in mathematics.

manifold
a fundamental mathematical object which locally resembles a line, a plane, or space (is locally homeomorphic to a vector space). For example, a sphere or a doughnut, the playing field of the video game Asteroids, and (in the theory of general relativity) physical space are all manifolds. The term n-manifold means a manifold which locally resembles n-dimensional space, not one which might lie in n-dimensional space.

manipulatives
a wide variety of physical materials and supplies that students use to foster the learning of abstract ideas in mathematics.

margin of error
a measure of the uncertainty in an estimate of a parameter; unfortunately, not everyone agrees what it should mean. The margin of error of an estimate is typically one or two times the esti-

mated standard error of the estimate.

Markov's inequality

if a list contains no negative numbers, the fraction of numbers in the list at least as large as any given constant a>0 is no larger than the arithmetic mean of the list, divided by a.

for random variables

if a random variable X must be nonnegative, the chance that X exceeds any given constant a>0 is no larger than the expected value of X, divided by a.

mass

the amount of matter in a body.

mathematical biology

the application of mathematics to problems in biology. Since mathematicians are biological entities, some mathematical biologists are motivated by the desire to write equations about themselves.

mathematical physics

the study of mathematical concepts used in physics, especially statistical mechanics and quantum field theory. Some say that modern mathematical physics is all areas of mathematics other than classical mathematical physics.

mathematics

the study of the relations between objects or quantities. These relations are organized so that certain facts can be derived from others by using logic. There are about 3000 categories of mathematics (e.g., applied, pure).

matrices

a rectangular array of numbers or letters arranged in rows and columns.

matrix model

a kind of quantum field theory in two dimensions involving matrix-valued fields which is related to random triangulations.

Matukuma's equation

the non-linear partial differential equation Nabla u + up/(1+v2) = 0, where u is a function of a vector v in n dimensions. The equation was proposed by Matukuma as a model for a spherical cluster of stars.

maxima

the points on a curve where the value is greater than that of the surrounding points.

maximum

the largest of a set of values.

maximum likelihood estimate (mle)

the maximum likelihood estimate of a parameter from data is the possible value of the parameter for which the chance of observing the data largest. That is, suppose that the parameter is p, and that we observe data x. Then the maximum likelihood estimate of p is estimate p by the value q that makes P(observing x when the value of p is q) as large as possible. For example, suppose we are trying to estimate the chance that a (possibly biased) coin lands heads when it is tossed. Our data will be the number of times x the coin lands heads in n independent tosses of the coin. The distribution of the number of times the coin lands heads is binomial with parameters n (known) and p (unknown). The chance of observing x heads in n trials if the chance of heads in a given trial is q is ${}^{n}C_{x}$ qx(1-q)n-x. The maximum likelihood estimate of p would be the value of q that makes that chance largest. We can find that value of q explicitly using calculus; it turns out to be q = x/n, the fraction of times the coin is observed to land heads in the n tosses. Thus the maximum likelihood estimate of the chance of heads from the number of heads in n independent tosses of the coin is the observed fraction of tosses in which the coin lands heads.

mean

1. the average of a set of numbers, calculated by dividing the sum by the number of values.

2. in statistics, the average obtained by dividing the sum of two or more quantities by the number of these quantities.

mean squared error (mse)

the mean squared error of an estimator of a parameter is the expected value of the square of the difference between the estimator and the parameter. In symbols, if X is an estimator of the parameter t, then.

mSE(X) = E((X-t)2).

the MSE measures how far the estimator is off from what it is trying to estimate, on the average in repeated experiments. It is a summary measure of the accuracy of the estimator. It combines any tendency of the estimator to overshoot or undershoot the truth (bias), and the variability of the estimator (SE). The MSE can be written in terms of the bias and SE of the estimator.

mSE(X) = (bias(X))2 + (SE(X))2.

mean, arithmetic mean

the sum of a list of numbers, divided by the number of numbers. See also average.

measures of central tendency

numbers that communicate the "center" or "middle" of a set of data. The mean, median and mode are statistical measures of central tendency.

mechanics.

Study of the forces acting on bodies, whether moving (dynamics) or stationary (statics).

medial triangle

the triangle whose vertices are the midpoints of the sides of a given triangle.

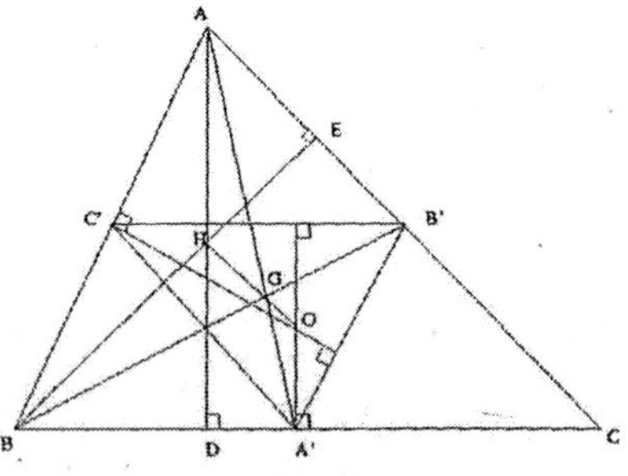

median

"middle value" of a list. The smallest number such that at

least half the numbers in the list are no greater than it. If the list has an odd number of entries, the median is the middle entry in the list after sorting the list into increasing order. If the list has an even number of entries, the median is the smaller of the two middle numbers after sorting. The median can be estimated from a histogram by finding the smallest number such that the area under the histogram to the left of that number is 50%.

■ member of a set

something is a member (or element) of a set if it is one of the things in the set.

■ mental math

computing an exact answer without using pencil and paper or other physical aids.

■ Mersenne prime

a Mersenne number, Mp, has the form 2p-1, where p is a prime. If Mp itself a prime, then it is called a Mersenne prime. There are 32 such primes known (i.e., not all primes yield a Mersenne prime). (see also Fermat prime.).

■ method of comparison

the most basic and important method of determining whether a treatment has an effect compare what happens to individuals who are treated (the treatment group) with what happens to individuals who are not treated (the control group).

■ metric

a distance function on some space or set; an assignment of distance to every unordered pair of points that satisfies the triangle inequality.

■ metric system

a system of measurements based on the decimal system, used in most countries and in almost all scientific applications.

■ metrisable

for a topological space, the property that there exists a metric compatible with the topology. To say that a topological space is metrisable is to treat it as a metric space, but without distinguishing any specific or preferred distance function.

■ **midpoint**

the point M is the medpoint of line segment AB if AM=MB. That is, M is halfway between A and B.

■ **min cut, max flow principle**

the principle that, given a flow of some substance through a network of pipes with a maximum allowed flow in each pipe, the maximum possible total flow equals the capacity of the most constricted set of pipes which, if cut, would separate the source from the drain.

■ **minima**

the points on a curve where the value is less than that of the surrounding points.

■ **minimal**

a widely used word in mathematics that generally means atomic or unsimplifiable. For example, prime numbers are minimal with respect to factorisation of whole numbers.

■ **minimal surface**

a surface (or manifold) which locally minimises surface area (or surface volume), which means that one cannot replace small patches of the surface and decrease the area.

■ **minimax strategy**

in game theory, a minimax strategy is one that minimizes one's maximum loss, whatever the opponent might do (whatever strategy the opponent might choose).

■ **minimum**

the smallest of a set of values.

■ **Minkowski space**

a finite-dimensional vector space, especially a 4-dimensional one, together with an indefinite inner product with one positive or timelike direction and many negative or spacelike directions. In particular, ordinary spacetime in the theory of Special Relativity.

■ **minor arc**
an arc on a circle that is less than 180 degrees.

■ **minor axis**
the shortest distance across an ellipse through the center.

■ **minuend**
the number from which another is to be subtracted. e.g., 7 - 5 = 2, 7 is the minuend.

■ **minute**
the unit of measure of an angle that is 1/60 of a degree.

■ **mira**
a reflective, plastic mirror used in geometry.

■ **mixed decimal**
a decimal number with an integer part and a decimal part.

■ **mixed fraction**
a number with an integer part and a fraction part.

■ **mixed number**
a whole number and a fraction, as in one-and-a-half (1 + 1/2).

■ **mode**
for lists, the mode is a most common (frequent) value. A list can have more than one mode. For histograms, a mode is a relative maximum ("bump").

■ **model (noun)**
a display of concrete materials, objects or drawings.

■ **model (verb)**
use of concrete materials, symbolic.

■ **moduli space**
given a topological object, usually low-dimensional, with a continuous family of possible geometric structures, the moduli space is the set of these structures considered itself as a geometric object, usually high-dimensional. Two structures that differ by a topological automorphism of the low-dimensional object are represented by the same point. In particular, the space of conformal structures on a surface.

■ **modulo**
the integers a and b are said to be congruent modulo m if a-b is divisible by m.

modulus

the absolute value of a complex number.

moment

the kth moment of a list is the average value of the elements raised to the kth power; that is, if the list consists of the N elements x1, x2, … , xN, the kth moment of the list is.

(x1k + x2k + xNk)/N.

the kth moment of a random variable X is the expected value of Xk, E(Xk).

monic polynomial

a polynomial in which the coefficient of the term of highest degree is 1.

monochromatic triangle

a triangle whose vertices are all colored the same.

monomial

in the variables x, y, z, a monomial is an expression of the form axmynzk, in which m, n, and k are non-negative integers and a is a constant (e.g., 5x2, 3x2y or 7x3yz2).

monotone, monotonic function

a function is monotone if it only increases or only decreases f increases monotonically (is monotonic increasing) if x > y, implies thatf(x) >= f(y). A function f decreases monotonically (is monotonic decreasing) if x > y, implies thatf(x) <= f(y). A function f is strictly monotonically increasing if x > y, implies thatf(x) > f(y), and strictly monotonically decreasing if if x > y, implies thatf(x) < f(y).

morphogenesis

the period of an embryo's existence where it determines the shape, organs, and body parts of the organism which it will become.

multimodal distribution

a distribution with more than one mode. The histogram of a multimodal distribution has more than one "bump."

multinomial

an algebraic expression consisting of 2 or more terms.

multinomial distribution

consider a sequence of n independent trials, each of which can result in an outcome in any of k categories. Let pj be the probability that each trial results in an outcome in category j, j = 1, 2, ... , k, so.

p1 + p2 + ... + pk = 100%.

the number of outcomes of each type has a multinomial distribution. In particular, the probability that the n trials result in n1 outcomes of type 1, n2 outcomes of type 2, ... , and nk outcomes of type k is.

n!/(n1! × n2! × ... × nk!) × p1n1 × p2n2 × ... × pknk,

if n1, ... , nk are nonnegative integers that sum to n; the chance is zero otherwise.

multiple

a number that is the product of a whole number and any other whole number. For example, a multiple of 4 is 8.

multiplicand

a number that is being multiplied by another number. A factor.

multiplication

the operation of repeated addition. For example, 4 x 3 is the same as 4 + 4 + 4. Four is used as an addend three times in 4 x 3.

multiplication rule

the chance that events A and B both occur (i.e., that event AB occurs), is the conditional probability that A occurs given that B occurs, times the unconditional probability that B occurs.

multiplicative identity

the number 1 is the multiplicative identity because 1 * a = a for all a.

multiplicative inverse

the number, b, that when multiplied by a number, a, gives a result of 1. Reciprocal. b = 1/a.

multiplicity in hypothesis tests

in hypothesis testing, if more than one hypothesis is tested, the actual significance level of the combined tests is not equal to the nominal significance level of the individual tests. See also false discovery rate.

■ **multivariate data**
a set of measurements of two or more variables per individual. See bivariate.

■ **mutually exclusive**
see disjoint events or disjoint sets.

■ **nadir**
the point on the celestial sphere in the direction downwards of the plumb-line.

■ **Nagel point**
in a triangle, the lines from the vertices to the points of contact of the opposite sides with the ex-circles to those sides meet in a point called the Nagel point.

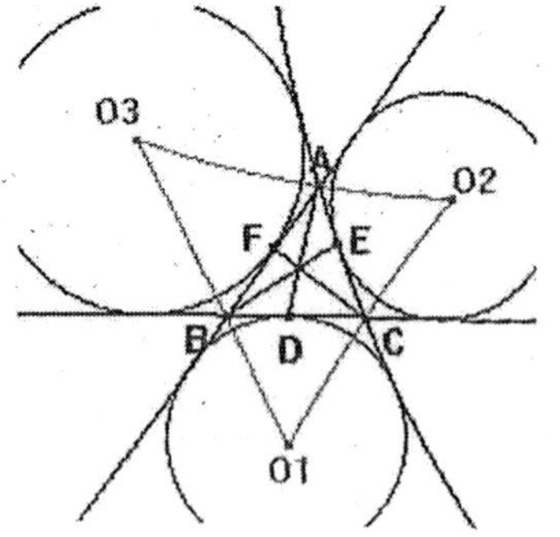

■ **natural logarithm**
any logarithm with Euler's number (e) as its base. A natural log is written as ln. For example, ln(e) = 1.

■ **natural numbers**
the counting numbers.

■ **Navier-Stokes equation**
the basic equation modelling the motion of a uniform fluid with no viscosity, but possibly varying density.

■ **nearly normal distribution**
a population of numbers (a list of numbers) is said to have a nearly normal distribution if the histogram of its values in standard units nearly follows a normal curve. More precisely, suppose that the mean of the list is μ and the standard deviation of the list is SD. Then the list is nearly normally distributed if, for every two numbers $a < b$, the fraction of numbers in the list that are between a and b is approximately equal to the area under the normal curve between $(a - \mu)/SD$ and $(a - \mu)/SD$.

negative binomial distribution

consider a sequence of independent trials with the same probability p of success in each trial. The number of trials up to and including the rth success has the negative Binomial distribution with parameters n and r. If the random variable N has the negative binomial distribution with parameters n and r, then.

p(N=k) = k-1Cr-1 × pr × (1-p)k-r,

for k = r, r+1, r+2, . . . , and zero for k < r, because there must be at least r trials to have r successes. The negative binomial distribution is derived as follows for the rth success to occur on the kth trial, there must have been r-1 successes in the first k-1 trials, and the kth trial must result in success. The chance of the former is the chance of r-1 successes in k-1 independent trials with the same probability of success in each trial, which, according to the Binomial distribution with parameters n=k-1 and p, has probability.

k-1Cr-1 × pr-1 × (1-p)k-r.

the chance of the latter event is p, by assumption. Because the trials are independent, we can find the chance that both events occur by multiplying their chances together, which gives the expression for P(N=k) above.

negative number

a real number less than zero.

nets

a plane figure obtained by opening and flattening a solid figure.

nine point centre

in a triangle, the circumcentre of the medial triangle is called the nine point centre.

nine point circle

in a triangle, the circle that passes through the midpoints of the sides is called the nine point circle.

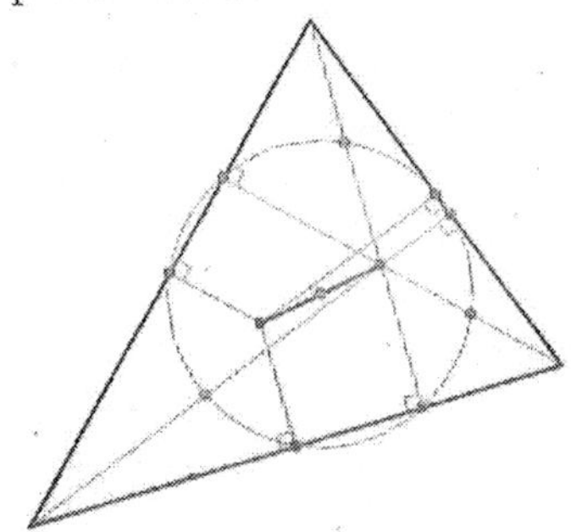

no causation without manipulation
a slogan attributed to Paul Holland. If the conditions were not deliberately manipulated (for example, if the situation is an observational study rather than an experiment), it is unwise to conclude that there is any causal relationship between the outcome and the conditions. See post hoc ergo propter hoc.

nomograph
a graphical device used for computation which uses a straight edge and several scales of numbers.

non-abelian
non-commutative or order-dependent. For example, the group of manipulations of the Rubik's cube is non-commutative because the state of the cube depends greatly on the order that moves are performed on it.

nonagonal number
a number of the form n(7n-5)/2.

nonanticipativity
a solution to a stochastic programming problem, in the form of a sequence of responses to random events, is non anticipative if each given responses depends only on past events and not on future events.

nonary
associated with 9

non-equilibrium statistical mechanics
the study of thermal and statistical properties of systems that are out of equilibrium, e.g., that are changing temperature or moving.

non-linear
referring to an equation, model, or solution, not amenable to linear algebra or having terms which are not strictly proportional to values of the variables.

nonlinear association
the relationship between two variables is nonlinear if a change in one is associated with a change in the other that is depends on the value

of the first; that is, if the change in the second is not simply proportional to the change in the first, independent of the value of the first variable.

■ **nonresponse**

in surveys, it is rare that everyone who is "invited" to participate (everyone whose phone number is called, everyone who is mailed a questionnaire, everyone an interviewer tries to stop on the street …) in fact responds. The difference between the "invited" sample sought, and that obtained, is the nonresponse.

■ **nonresponse bias**

in a survey, those who respond may differ from those who do not, in ways that are related to the effect one is trying to measure. For example, a telephone survey of how many hours people work is likely to miss people who are working late, and are therefore not at home to answer the phone. When that happens, the survey may suffer from nonresponse bias. Nonresponse bias makes the result of a survey differ systematically from the truth.

■ **nonresponse rate**

the fraction of nonresponders in a survey the number of nonresponders divided by the number of people invited to participate (the number sent questionnaires, the number of interview attempts, etc.) If the nonresponse rate is appreciable, the survey suffer from large nonresponse bias.

■ **nonstandard measurement**

measurement expressed in terms of objects such as paper clips, sticks of gun, shoes, etc.

■ **nonstandard unit**

unit of measurement expressed in terms of objects (such as paper clips, sticks of gum, shoes, etc.).

■ **non-standard unit**

measurement expressed in terms of objects such as paper clips, sticks of gum, shoes, etc.

norm

(1) mean; average. (2) customary degree or condition. (3) established pattern or form.

normal

perpendicular.

normal approximation

the normal approximation to data is to approximate areas under the histogram of data, transformed into standard units, by the corresponding areas under the normal curve. Many probability distributions can be approximated by a normal distribution, in the sense that the area under the probability histogram is close to the area under a corresponding part of the normal curve. To find the corresponding part of the normal curve, the range must be converted to standard units, by subtracting the expected value and dividing by the standard error. For example, the area under the binomial probability histogram for n = 50 and p = 30% between 9.5 and 17.5 is 74.2%. To use the normal approximation, we transform the endpoints to standard units, by subtracting the expected value (for the Binomial random variable, n×p = 15 for these values of n and p) and dividing the result by the standard error (for a Binomial, (n × p × (1-p))1/2 = 3.24 for these values of r and p). The area normal approximation is the area under the normal curve between (9.5 - 15)/3.24 = -1.697 and (17.5 - 15)/3.24 = 0.772; that area is 73.5% slightly smaller than the corresponding area under the binomial histogram. See also the continuity correction. The tool on this page illustrates the normal approximation to the binomial probability histogram. Note that the approximation gets worse when p gets close to 0 or 1, and that the approximation improves as n increases.

normal curve

the normal curve is the familiar "bell curve," illustrated on this page. The mathematical expression for the normal curve is y = (2×pi)-½E-x2/2, where pi is the ratio of the circumference of a circle to its diameter (3.14159265 ...), and E is the base of the natural logarithm (2.71828 ...). The normal curve is symmetric around the point x=0, and positive for every value of x. The area under the normal curve is unity, and the SD of the normal curve, suitably defined, is also unity. Many (but not most) histograms, converted into standard units, approximately follow the normal curve.

normal distribution

a random variable X has a normal distribution with mean m and standard error s if for every pair of numbers a <= b, the chance that a < (X-m)/s < b is P(a < (X-m)/s < b) = area under the normal curve between a and b.If there are numbers m and s such that X has a normal distribution with mean m and standard error s, then X is said to have a normal distribution or to be normally distributed. If X has a normal distribution with mean m=0 and standard error s=1, then X is said to have a standard normal distribution. The notation X~N(m,s2) means that X has a normal distribution with mean m and standard error s; for example, X~N(0,1), means X has a standard normal distribution.

not, negation, logical negation

the negation of a logical proposition p, NOT p, is a proposition that is the logical opposite of p. That is, if p is true, NOT p is false, and if p is false, NOT p is true. Negation takes precedence over other logical operations.

nucleation

a process in a physical system, or a mathematical

model such as a cellular automaton or a statistical model, whereby a bubble or other structure appears spontaneously at a random or unpredictable spot.

■ **null hypothesis**
in hypothesis testing, the hypothesis we wish to falsify on the basis of the data. The null hypothesis is typically that something is not present, that there is no effect, or that there is no difference between treatment and control.

■ **null set**
the empty set.

■ **number line**
a line on which each point represents a real number.

■ **number theory**
the study of properties of integers, generalisations of integers, and relations between them, especially Diophantine equations (equations in integers) and prime numbers.

■ **numeral**
a symbol or name that represents a number. e.g., three, 3.

■ **numerator**
the quantity above the line in a fraction. It tells how many parts are being counted.

■ **numerical analysis**
the study of methods for approximation of solutions of various classes of mathematical problems including error analysis.

■ **oblate spheroid**
an ellipsoid produced by rotating an ellipse through 360° about its minor axis.

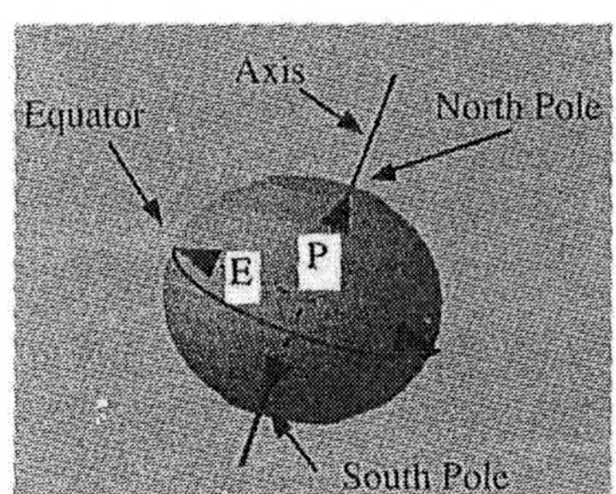

■ **oblique angle**
an angle which is not a right angle, can be obtuse or acute.

■ **oblique coordinates**
a coordinate system in which the axes are not perpendicular.

■ **oblique triangle**
a triangle that is not a right triangle.

■ observational study

c.f. controlled experiment.

■ obtuse angle

an angle with a degree measure between 90 and 180.

■ obtuse triangle

a triangle with one angle greater than 90 degrees, and therefore the other two angles must be acute.

■ octagon

a polygon with 8 sides.

■ octahedron

a polyhedron with 8 faces.

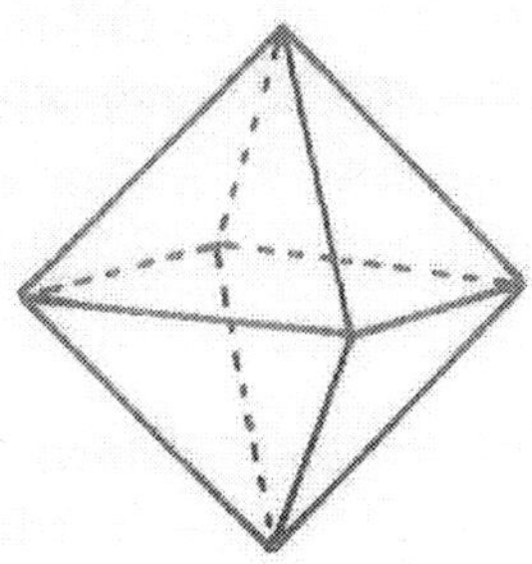

■ octal number

a number in base 8.

■ octant

any one of the 8 portions of space dtermined by the 3 coordinate planes.

■ odd function

a function that satisfies the property that f(-x) = -f(x).

■ odd number

a natural number that is not divisible by 2.

■ odds

the odds in favor of an event is the ratio of the probability that the event occurs to the probability that the event does not occur. For example, suppose an experiment can result in any of n possible outcomes, all equally likely, and that k of the outcomes result in a "win" and n-k result in a "loss." Then the chance of winning is k/n; the chance of not winning is (n-k)/n; and the odds in favor of winning are (k/n)/((n-k)/n) = k/(n-k), which is the number of favorable outcomes divided by the number of unfavorable outcomes. Note that odds are not synonymous with probability, but the two can be converted back and forth. If the odds in favor of an event are q, then the probability of the event is q/(1+q). If the probability of an event is p, the odds in favor of the event are p/(1-p) and the odds against the event are (1-p)/p.

■ one to one

a function f is said to be one to one if f(x)=f(y) implies that x=y.

■ onto

a function f is said to map A onto B if for every b in B, there is some a in A such f(a)=b.

■ open interval

an interval that does not include its two endpoints.

■ open sentence

a statement that contains at least one unknown. For example, 6 + x = 14.

■ optimal control

the study of problems that combine optimisation with control theory; how to set the behaviour machines so that they respond as efficiently as possible to variable or time-dependent conditions.

■ oR, disjunction, logical disjunction

an operation on two logical propositions. If p and q are two propositions, (p OR) q is a proposition that is true if p is true or if q is true (or both); otherwise, it is false. That is, (p OR) is true unless both p and q are false. C.f. exclusive disjunction, XOR.

■ orbifold

a topological object defined by Thurston which is locally modelled by Euclidean spaces divided by finite groups of symmetries. Orbifolds are manifolds with singularities such as reflection surfaces, where they resemble manifolds with boundary, and cone lines, where they are modelled (in the direction perpendicular to the cone line) by a cone with an angle of 360/n degrees for some n.

■ order of operations

the specific order in which mathematical operations are to be performed.

■ ordered pair

1. a set of two numbers in which the order has an agreed upon meaning. Such as the cartesian coordinates (x, y), where it is agreed that the first coordinate represents the horizontal position, and the

second coordinate represents the vertical position.
2. two numbers which give the "address" of a point in a plane.
e.g., (2, 5).

ordinal numbers
numbers showing order or position.
e.g., first, second, etc.

ordinal variable
a variable whose possible values have a natural order, such as {short, medium, long}, {cold, warm, hot}, or {0, 1, 2, 3, . . . }. In contrast, a variable whose possible values are {straight, curly} or {Arizona, California, Montana, New York} would not naturally be ordinal. Arithmetic with the possible values of an ordinal variable does not necessarily make sense, but it does make sense to say that one possible value is larger than another.

ordinary differential equation/ODE
any equation relating a function of one variable to its derivatives.

ordinate
the vertical coordinate on a plane.

origin
the point on a graph that represents the point where the x and y axes meet (x,y) = (0,0).

orthic triangle
the triangle whose vertices are the feet of the altitudes of a given triangle.

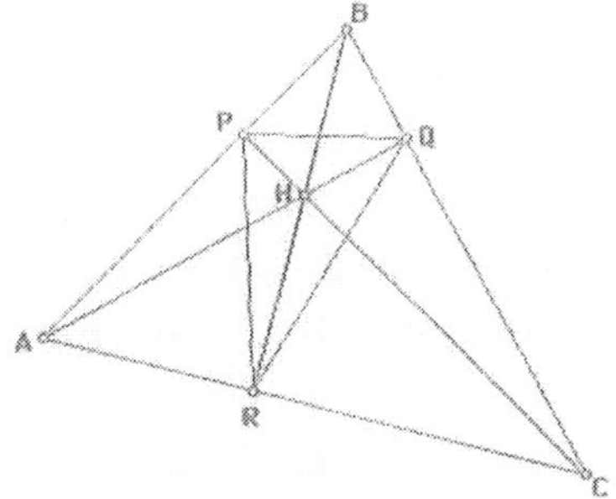

orthocentre
the point of intersection of the altitudes of a triangle.

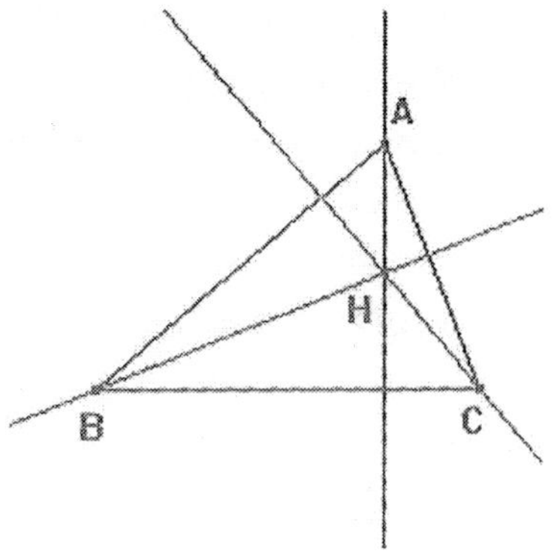

■ **orthogonal**

perpendicular.

■ **outcome space**

the outcome space is the set of all possible outcomes of a given random experiment. The outcome space is often denoted by the capital letter S.

■ **outlier**

an outlier is an observation that is many SD's from the mean. It is sometimes tempting to discard outliers, but this is imprudent unless the cause of the outlier can be identified, and the outlier is determined to be spurious. Otherwise, discarding outliers can cause one to underestimate the true variability of the measurement process.

■ **Painlevé equation/ Painlevé function**

the Painlevé equations are six families (denoted by Roman numerals) of second-order ODEs whose solutions have relatively simple behaviour in the complex plane. (Namely, their branch points as complex functions depend only on the equation and not on the particular solution) The solutions are called Painlevé functions. The simplest such equation (Painlevé I) is y'' = 6y^2 + x.

■ **palindrome**

a positive integer whose digits read the same forward and backwards.

■ **palindromic**

a positive integer is said to be palindromic with respect to a base b if its representation in base b reads the same from left to right as from right to left.

■ **pandiagonal magic square**

a magic square in which all the broken diagonals as well as the main diagonals add up to the magic constant.

■ **pandigital**

a decimal integer is called pandigital if it contains each of the digits from 0 to 9.

■ **parabola**

1. the set of all points in a plane that are equally distant from a fixed point (called the

focus) and a fixed line, (called the directrix).

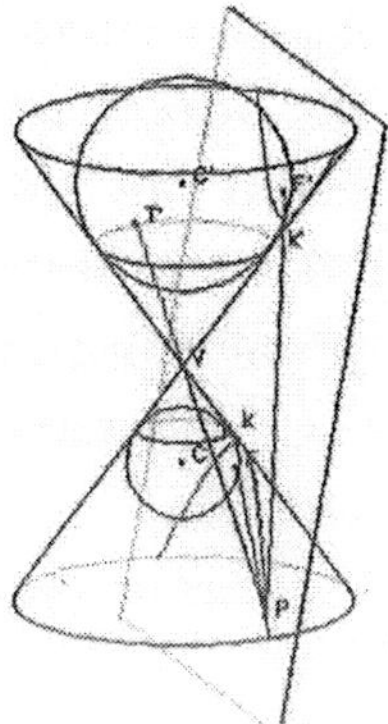

2. the set of points equidistance from a focus and a directrix.

■ **paraboloid**

1. a paraboloid of revolution is a surface of revolution produced by rotating a parabola about its axis.
2. a surface that is formed by rotating a parabola about its axis.

■ **parallel**

given distinct lines in the plane that are infinite in both directions, the lines are parallel if they never meet. Two distinct lines in the coordinate plane are parallel if and only if they have the same slope.

■ **parallelepiped**

a solid figure with six faces such that the planes containing two opposite faces are parallel. Each face is a parallelogram.

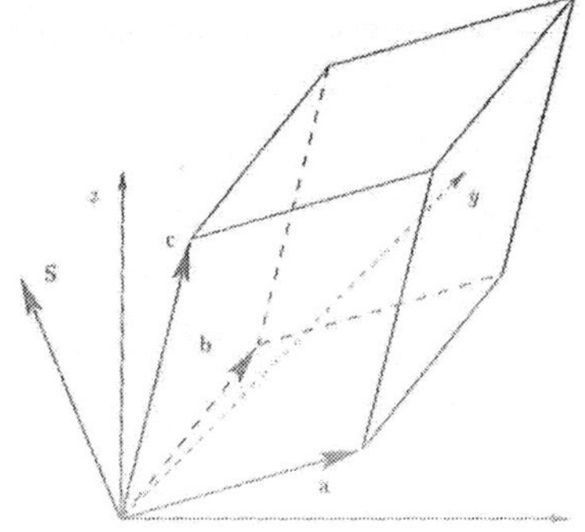

■ **parallelism**

the state of being parallel, not intersecting.

■ **parallelogram**

a quadrilateral with opposite sides parallel.

■ **parameter**

a quantity whose value varies with the circumstances of its application, such as the radius of a group of circles.

■ **partial differential equation/PDE**

any equation relating a function of several variables to

its partial derivatives in different directions.

partition

a partition of an event B is a collection of events {A1, A2, A3, … } such that the events in the collection are disjoint, and their union is B (they exhaust B). That is, AjAk = {} unless j = k, and B = A1 U A2 U A3 U …. If the event B is not specified, it is assumed to be the entire outcome space S.

part-part whole mat

a form used to assist in recognising number relationships.

Pascal's triangle

a triangular array of binomial coefficients.

payoff matrix

a way of representing what each player in a game wins or loses, as a function of his and his opponent's strategies.

Peano axioms

the axiom system developed by Giuseppe Peano to formalise arithmetic. The key to his method is the introduction of a "successor operation" S on numbers; if a is a number, then Sa is the successor of that number. In this way Peano reduced arithmetic to the conceptually primitive operation of counting.

0 is a number.

if a is a number, then Sa is a number.

if a and b are numbers, then a + 0 = a, and a + Sb = S(a + b).

if a and b are numbers, then a × 0 = 0, and a × Sb = a × b + a.

Peano, Giuseppe (born 1858)

italian mathematician best known for the Peano axioms, which define the natural numbers in terms of sets and formalise arithmetic in a logically rigorous way. Peano was primarily an analyst, however, and was the inventor of space-filling curves, that is surjective functions from the unit interval to the unit square.

pedal triangle

the pedal triangle of a point P with respect to a triangle ABC is the triangle whose vertices are the feet of the perpendiculars dropped from P to the sides of triangle ABC.

Pell number

the nth term in the sequence 0, 1, 2, 5, 12,... defined by the recurrence.

p0=0, P1=1, and Pn=2Pn-1+Pn-2.

pentagon

a five sided polygon.

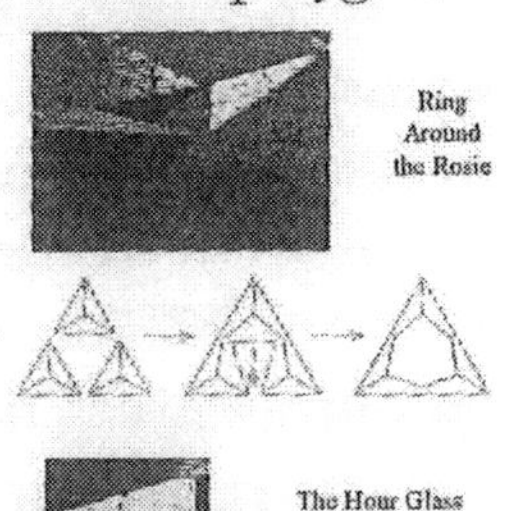

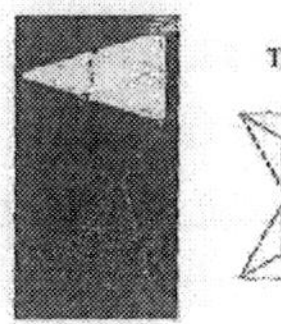

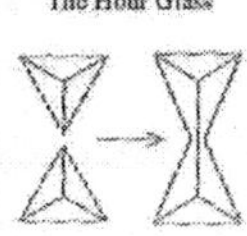

pentagonal number

a number of the form n(3n-1)/2.

pentomino

a five-square polyomino.

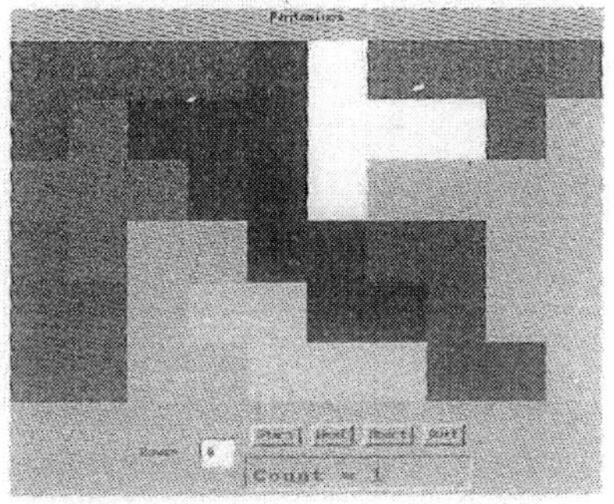

percent

the amount out of 100. 4/5 is equal to 80/100, and is thus 80%.

percentile

the pth percentile of a list is the smallest number such that at least p% of the numbers in the list are no larger than it. The pth percentile of a random variable is the smallest number such that the chance that the random variable is no larger than it is at least p%. C.f. quantile.

perfect cube
an integer is a perfect cube if it is of the form m3 where m is an integer.

perfect number
a number that is the sum of all its factors except itself. For example, 6.

perfect number
a number which is equal to the sum of its proper divisors. 6, 28, and 496 are the three of seven known perfect numbers. [6 is a perfect number because its proper divisors (1,2, and 3) total 6.]

perfect power
an integer is a perfect power if it is of the form mn where m and n are integers and n>1.

perfect square
an integer is a perfect square if it is of the form m2 where m is an integer.

perimeter
the sum of the lengths of the sides of a polygon.

periodic function
a function that keeps repeating the same values.

periodic trajectory
a trajectory of a flow that repeats, or makes a topological circle.

permanent
a function of a matrix which has the same terms as the determinant, but all terms are added and none are subtracted.

permanent-determinant method
a scheme, found by P. W. Kasteleyn, to change the signs of the entries of a matrix with planar sparseness in such a way that the permanent becomes the determinant.

permutation
a permutation of a set is an arrangement of the elements of the set in some order. If the set has n things in it, there are n! different orderings of its elements. For the first element in an ordering, there are n possible choices, for the second, there remain n-1 possible choices, for the third, there are n-2, etc., and for the nth element of the

ordering, there is a single choice remaining. By the fundamental rule of counting, the total number of sequences is thus n×(n-1)×(n-2)× … ×1. Similarly, the number of orderings of length k one can form from n>=k things is n×(n-1)×(n-2)× … ×(n-k+1) = n!/(n-k)!. This is denoted nPk, the number of permutations of n things taken k at a time. C.f. combinations.

■ **perpendicular**
at right angles to a line or plane.

■ **pi (π)**
the ratio of the circumference of a circle to its diameter. The value of p is 3.1415926, correct to seven decimal places.

■ **pie plates**
two pie plates (waxed) with a cut from the outer part of the plate to the centre and slid together so as to rotate on each other. Used to develop the concept of fractions.

■ **piecewise defined function**
using different rules for different parts of the domain.

■ **PL flow**
a motion on a space or a manifold, akin to a flow given by a vector field, in which every particle in a given simplex of some triangulation moves with constant velocity and in the same direction, so that the particle trajectories are polygons.

■ **place value**
the value of the position of a digit in a number. For example, in the number 7863, the 8 is in the hundreds place and its value is 800.

■ **placebo**
a "dummy" treatment that has no pharmacological effect; e.g., a sugar pill.

■ **placebo effect**
the belief or knowledge that one is being treated can itself have an effect that confounds with the real effect of the treatment. Subjects given a placebo as a pain-killer report statisti-

cally significant reductions in pain in randomised experiments that compare them with subjects who receive no treatment at all. This very real psychological effect of a placebo, which has no direct biochemical effect, is called the placebo effect. Administering a placebo to the control group is thus important in experiments with human subjects; this is the essence of a blind experiment.

■ planar sparseness

a matrix M has planar sparseness if the graph formed by connecting row i to column j whenever M_ij is non-zero is planar.

■ plane

a flat surfaces that stretches off into infinity.

■ plane field

a decoration of a n-manifold that assigns a tangent k-dimensional plane to every point in a continuous fashion. The general notion includes line fields and hyperplane fields; line fields are very similar to vector fields such as the magnetic field.

■ plane geometry

the study of two-dimensional figures.

■ plane partition

a stack of unit cubes in a rectangular box or in the positive octant in space such that to the left, behind, and below every cube lies either another cube or a wall. A plane partition in a box is equivalent to a lozenge tiling of a hexagon in the plane.

■ Platonism

the belief that mathematical objects exist independent of physical models. It is a useful pretense in mathematics, especially in geometry.

■ pleated surface

a surface in Euclidean or hyperbolic space which resembles a polyhedron in the sense that it has flat faces that meet along edges. Unlike a polyhedron, a pleated surface has no corners, but it may have infinitely many edges that form a lamination.

Poincaré conjecture

the conjecture that a closed, simply-connected 3-manifold must be homeomorphic to the 3-sphere. Many mathematicians, including Poincaré himself, have presented incomplete and incorrect proofs of the conjecture.

point

a point in geometry is undefined. A geometric point is considered to be a particular location in geometric space. A point does not have length, width, or depth.

point of averages

in a scatterplot, the point whose coordinates are the arithmetic means of the corresponding variables. For example, if the variable X is plotted on the horizontal axis and the variable Y is plotted on the vertical axis, the point of averages has coordinates (mean of X, mean of Y).

Poisson distribution

the Poisson distribution is a discrete probability distribution that depends on one parameter, m. If X is a random variable with the Poisson distribution with parameter m, then the probability that X = k is E-m × mk/k!, k = 0, 1, 2, . . . ; where E is the base of the natural logarithm and ! is the factorial function. For all other values of k, the probability is zero. The expected value the Poisson distribution with parameter m is m, and the standard error of the Poisson distribution with parameter m is $m^{1/2}$.

polar coordinates

the coordinate system for the plane based on r, ?, the distance from the origin and ?, and the angle between the positive x-axis and the ray from the origin to the point.

polar equations

any relation between the polar coordinates (r,Ø) of a set of points. (e.g., r= 2cosØ is the polar equation of a circle referred to a point on its circumference and its diameter as an initial line).

polygon

a geometric figure that is bound by many straight lines such as triangle, square, pentagon, hexagon, heptagon, octagon etc.

polyhedron

1. A region in Euclidean space which consists of flat facets with flat edges. More technically, a polyhedron must locally be a cone over a lower-dimensional polyhedron. It is sometimes but not always implicitly assumed that a polyhedron is a manifold, a topological sphere or ball, or a convex set.

2. An abstract space with properties analogous to that of a polyhedron, such as a simplicial complex.

polynomial

in algebra, an expression consisting of two or more terms, such as x2 - 2xy + y2

polyomino

a planar figure consisting of congruent squares joined edge-to-edge.

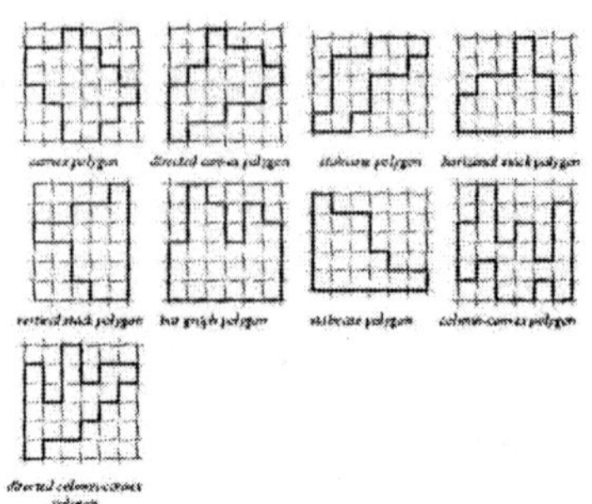

polytope

the n-dimensional generalisation of a polygon or a polyhedron.

population

a collection of units being studied. Units can be people, places, objects, epochs, drugs, procedures, or many other things. Much of statistics is concerned with estimating numerical properties (parameters) of an entire population from a random sample of units from the population.

population mean

the mean of the numbers in a numerical population. For example, the population mean of a box of numbered tickets is the mean of the list comprised of all the numbers on all the tickets. The population mean is a parameter. C.f. sample mean.

■ population percentage

the percentage of units in a population that possess a specified property. For example, the percentage of a given collection of registered voters who are registered as Republicans. If each unit that possesses the property is labeled with "1," and each unit that does not possess the property is labeled with "0," the population percentage is the same as the mean of that list of zeros and ones; that is, the population percentage is the population mean for a population of zeros and ones. The population percentage is a parameter. C.f. sample percentage.

■ population standard deviation

the standard deviation of the values of a variable for a population. This is a parameter, not a statistic. C.f. sample standard deviation.

■ positive number

a real number greater than zero.

■ post hoc ergo propter hoc

"After this, therefore because of this." A fallacy of logic known since classical times inferring a causal relation from correlation. Don't do this at home!

■ postulate

a fundamental statement that is assumed to be true without proof.

■ power

refers to an hypothesis test. The power of a test against a specific alternative hypothesis is the chance that the test correctly rejects the null hypothesis when the alternative hypothesis is true.

■ practical number

a practical number is a positive integer m such that every natural number n not exceeding m is a sum of distinct divisors of m.

prima facie
latin for "at first glance." "On the face of it." Prima facie evidence for something is information that at first glance supports the conclusion. On closer examination, that might not be true; there could be another explanation for the evidence.

prime
a natural number p greater than 1 is prime if and only if the only positive integer factors of p are 1 and p. The first seven primes are 2, 3, 5, 7, 11, 13, 17.

prime factorisation
the calculation of every prime factor in a given number. The prime factors of 12 are 3, 2, and 2.

prime factors
Prime factors of a number are a list of prime numbers the product of which is the number concerned. When n=1, for example, f(x)=2x1+3, this is a linear expression. If n=2, it is quadratic (for example, x2 + 2x + 4); if n=3, it is cubic, if n=4, it is quartic and if n=5, it is quintic.

prime number
a natural number other than 1, evenly divisible only by 1 and itself. The numbers 2,3,5,7,11,13,17,19,... Apart from 2 all primes are odd numbers and odd primes fall into two groups those that are one less than a multiple of four (3,7,11,19) and those one more than a multiple of four (5,13,17). Every natural number greater than 1 may be resolved into a product of prime numbers; eg 8316 = 22 x 33 x 7 x 11.

primes
counting numbers that can only be evenly divided by two numbers which are the number itself and 1. For example, the numbers 2, 3, 5, 7.

primitive Pythagorean triangle
a right triangle whose sides are relatively prime integers.

primitive root of unity
the complex number z is a primitive nth root of unity if

zn=1 but zk is not equal to 1 for any positive integer k less than n.

prism

a solid with parallel bases, congruent polygons, and sides that are all parallelograms.

probability

1. the odds of an event occuring. 50% means the event will occur 1 out of every 2 times

2. the probability of an event is a number between zero and 100%. The meaning (interpretation) of probability is the subject of theories of probability, which differ in their interpretations. However, any rule for assigning probabilities to events has to satisfy the axioms of probability.

probability density function

the chance that a continuous random variable is in any range of values can be calculated as the area under a curve over that range of values. The curve is the probability density function of the random variable. That is, if X is a continuous random variable, there is a function f(x) such that for every pair of numbers a<=b, P(a<= X <=b) = (area under f between a and b); f is the probability density function of X. For example, the probability density function of a random variable with a standard normal distribution is the normal curve. Only continuous random variables have probability density functions.

probability histogram

a probability histogram for a random variable is analogous to a histogram of data, but instead of plotting the area of the bins proportional to the relative frequency of observations in the class interval, one plots the area of the bins proportional to the probability that the random variable is in the class interval.

■ **probability sample**

a sample drawn from a population using a random mechanism so that every element of the population has a known chance of ending up in the sample.

■ **probability, theories of**

a theory of probability is a way of assigning meaning to probability statements such as "the chance that a thumbtack lands point-up is 2/3." That is, a theory of probability connects the mathematics of probability, which is the set of consequences of the axioms of probability, with the real world of observation and experiment. There are several common theories of probability. According to the frequency theory of probability, the probability of an event is the limit of the percentage of times that the event occurs in repeated, independent trials under essentially the same circumstances. According to the subjective theory of probability, a probability is a number that measures how strongly we believe an event will occur. The number is on a scale of 0% to 100%, with 0% indicating that we are completely sure it won't occur, and 100% indicating that we are completely sure that it will occur. According to the theory of equally likely outcomes, if an experiment has n possible outcomes, and (for example, by symmetry) there is no reason that any of the n possible outcomes should occur preferentially to any of the others, then the chance of each outcome is 100%/n. Each of these theories has its limitations, its proponents, and its detractors.

■ **product**

the answer to a multiplication question.

e.g., 2 x 5 = 10, 10 is the product.

progressive hedging
an algorithm in stochastic programming to find the optimal sequence of responses to random events. For each response, the algorithm progressively stiffens penalties for dependence on future events to enforce nonanticipativity constraints.

projective plane
a 2-manifold obtained by gluing a Möbius strip and a disk along their circle boundaries.

pronic number
a number of the form n(n+1).

proper divisor
the integer d is a proper divisor of the integer n if 0<d<n and d is a divisor of n.

proper factor
the factors of a number except itself. The proper factors of 12 are 1,2,3,4 and 6.

proper fraction
a fraction whose numerator is an integer less than its integer denominator. For example, 4/5 is a proper fraction.

proportion
an equation showing that two ratios are equivalent. For example, 1/2 = 2/4 .

proposition, logical proposition
a logical proposition is a statement that can be either true or false. For example, "the sun is shining in Berkeley right now" is a proposition. See also AND, IFF, IMPLIES, OR, XOR, converse, and contrapositive.

protractor
a device for measuring angles.

prym variety
an algebraic variety associated to a conformal map between two Riemann surfaces. Like the Jacobian variety, it is a quotient of a certain complex vector space by a lattice; in particular, if the target Riemann surface is the Riemann sphere, it is the Jacobian variety.

■ pure mathematics

mathematics for the sake of its internal beauty or logical strength.

■ p-value

suppose we have a family of hypothesis tests of a null hypothesis that let us test the hypothesis at any significance level p between 0 and 100% we choose. The P value of the null hypothesis given the data is the smallest significance level p for which any of the tests would have rejected the null hypothesis. For example, let X be a test statistic, and for p between 0 and 100%, let xp be the smallest number such that, under the null hypothesis, $P(X <= x) >= p$. Then for any p between 0 and 100%, the rule reject the null hypothesis if $X < xp$ tests the null hypothesis at significance level p. If we observed $X = x$, the P-value of the null hypothesis given the data would be the smallest p such that $x < xp$.

■ pyramid

the union of all line segments that connect a given point and the points that lie on a given polygon.

■ Pythagoras (born 580 BC)

what we know of this ancient Greek mystic is mostly legend, for of the true facts of his life almost nothing is known. However, it is believed that he started a religious cult, the Pythagoreans, among whose beliefs was that number was the highest reality, and that all other aspects of reality could be understood in terms of integers and ratios of integers.

■ **Pythagorean Theorem**

1. the theorem that relates the three sides of a right triangle.
$a^2 + b^2 = c^2$
2. in Euclidean geometry, the sum of the areas of the squares on the legs of any right triangle is equal to the area of the square on the hypotenuse. This is arguably the most important theorem of classical mathematics, and perhaps of all time.

■ **Pythagorean triangle**

a right triangle whose sides are integers.

■ **Pythagorean triple**

an ordered set of three positive integers (a,b,c) such that $a^2+b^2=c^2$.

■ **quadrangle**

a closed broken line in the plane consisting of 4 line segments.

■ **quadrangular prism**

a prism whose base is a quadrilateral.

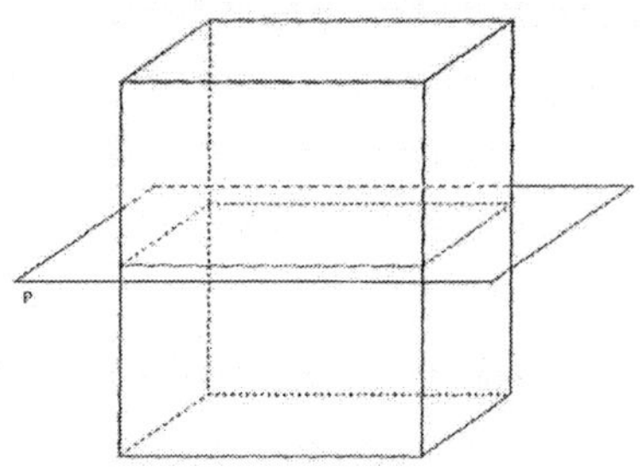

■ **quadrangular pyramid**

a pyramid whose base is a quadrilateral.

■ **quadrant**

1.any one of the four portions of the plane into which the plane is divided by the coordinate axes.
2. each of the four regions created by the coordinate axis.

■ **quadratfree**

square free

■ **quadratic equation**

an equation involving the second power, but no higher power of an unknown. The general form of a quadratic equation in two unknowns is.
$ax^2 + Bxy + Cy^2 + Dx + Ey + F = 0.$

quadratic equation
an equation of the form f(x)=0 where f(x) is a second degree polynomial. That is, $ax^2+bx+c=0$.

quadratic formula
the formula that says that the solution to a 2nd degree (quadratic) equation is as follows.

quadratic function
a function that has an equation of the form $y= Ax^2 + Bx + C$ where "A" does not equal 0.

quadratic function
a function given by a polynomial of degree 2.

quadrature
the quadrature of a geometric figure is the determination of its area.

quadric curve
the graph of a second degree equation in two variables.

quadric surface
the graph of a second degree equation in three variables.

quadrilateral
a polygon having four sides. Examples of quadrilaterals are squares, rectangles, parallelograms, and trapezoids.

quadrinomial
an algebraic expression consisting of 4 terms.

qualitative variable
a qualitative variable is one whose values are adjectives, such as colors, genders, nationalities, etc. C.f. quantitative variable and categorical variable.

quantile
the qth quantile of a list ($0 < q <= 1$) is the smallest number such that the fraction q or more of the elements of the list are less than or equal to it. i.e., if the list contains n numbers, the qth quantile, is the smallest number Q such that at least $n \times q$ elements of the list are less than or equal to Q.

quantitative variable
a variable that takes numerical values for which arithmetic makes sense, for ex-

ample, counts, temperatures, weights, amounts of money, etc. For some variables that take numerical values, arithmetic with those values does not make sense; such variables are not quantitative. For example, adding and subtracting social security numbers does not make sense. Quantitative variables typically have units of measurement, such as inches, people, or pounds.

quantum field theory

the study of force fields such as the electromagnetic field in the context of quantum mechanics, and often special relativity also. The mainstay of modern high-energy physics.

quantum G_2 link invariant

a polynomial invariant of knots and links, similar to the Jones polynomial, and associated with the Lie group or Lie algebra G_2, in particular its quantum group.

quantum gravity

the study of general relativity as a quantum field theory; the theory of integration of quantities over the space of Riemannian or Minkowskian metrics on a manifold.

quantum group

a generalisation of a group which has an associative multiplication law but does not have group elements in the usual sense. As a space, a quantum group is defined by non-commuting operators which are analogous to coordinates on a Lie group in the same way that non-commuting operators represent measurable quantities in quantum mechanics.

quantum mechanics

a theory of physics that says that all phenomena in the universe follow non-standard probabilistic laws called the quantum rules. The quantum rules radically affect the behaviour of any physical system with few available states, such as atoms and elementary particles.

■ **quartic**
a polynomial of degree 4.

■ **quartic polynomial**
a polynomial of degree 4.

■ **quartile**
there are three quartiles. The first or lower quartile (LQ) of a list is a number (not necessarily a number in the list) such that at least 1/4 of the numbers in the list are no larger than it, and at least 3/4 of the numbers in the list are no smaller than it. The second quartile is the median. The third or upper quartile (UQ) is a number such that at least 3/4 of the entries in the list are no larger than it, and at least 1/4 of the numbers in the list are no smaller than it. To find the quartiles, first sort the list into increasing order. Find the smallest integer that is at least as big as the number of entries in the list divided by four. Call that integer k. The kth element of the sorted list is the lower quartile. Find the smallest integer that is at least as big as the number of entries in the list divided by two. Call that integer l. The lth element of the sorted list is the median. Find the smallest integer that is at least as large as the number of entries in the list times 3/4. Call that integer m. The mth element of the sorted list is the upper quartile.

■ **quintic**
a polynomial of degree 5.

■ **quintic polynomial**
a polynomial of degree 5.

■ **quotient**
the answer to a division question.
e.g., 12,2 = 6, 6 is the quotient.

■ **R**
abbreviation for the real numbers.

■ **radian**
a unit of angular measurement such that there are 2 π radians in a complete circle. One radian = 180/π degrees. One radian is approximately 57.3°.

■ **radical**
the square root sign.

radical axis

the locus of points of equal power with respect to two circle.

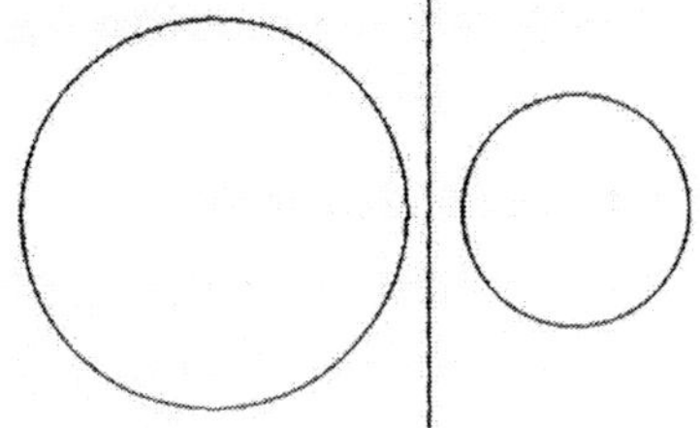

radical centre

the radical centre of three circles is the common point of interesection of the radical axes of each pair of circles.

radius

the distance from the center to a point on a circle. The line segment from the center to a point on a circle.

radix point

the generalisation of decimal point to bases of numeration other than base 10.

Ramsey's Theorem

given any two positive integers m and n, there exists a number, called the Ramsey number of m and n, and denoted R(m, n), such that any simple graph with R(m, n) vertices either contains a clique with m vertices (all vertices adjacent) or an independent set with n vertices (no vertices adjacent).

the question of how to determine the Ramsey number for arbitrary m and n is unsolved and believed to be very difficult, but various lower and upper bounds on it have been proven over the years. In particular, it is known that R(m, m) (≥) 2m/2, and that R(m, n) (≤) (m + n - z) Cm–1).

also, certain specific Ramsey numbers are known, e.g., R (3, 3) = 6. Thus, in any group of six people, either three are mutual acquaintances or three are mutual strangers.

■

random error

all measurements are subject to error, which can often be broken down into two components a bias or systematic error, which affects all measurements the same way; and a random error, which is in general different each time a measurement is made, and behaves like a number drawn with replacement from a box of numbered tickets whose average is zero.

random event

see random experiment.

random experiment

an experiment or trial whose outcome is not perfectly predictable, but for which the long-run relative frequency of outcomes of different types in repeated trials is predictable. Note that "random" is different from "haphazard," which does not necessarily imply long-term regularity.

random matrix

a matrix whose entries are random variables, often statistically independent. One then considers such questions as the likely eigenvalues or expected determinant of such a matrix.

random matrix theory

the study of matrices chosen at random from some specific probability distribution, especially real, complex, or quaternionic (symplectic) matrices whose entries are independent Gaussian random variables.

random number

a number that is supposedly chosen with no better odds than any other number. It is impossible to generate a true random number, but a computer can come close.

random sample

a random sample is a sample whose members are chosen at random from a given population in such a way that the chance of obtaining any particular sample can be computed. The number of units in the sample is called the sample size, often denoted n. The number of units in the population often is denoted N. Random samples can be drawn with or without replacing objects between draws; that is, drawing all n objects in the sample at once (a random sample without replacement), or drawing the objects one at a time, replacing them in the population between draws (a random sample with replacement). In a random sample with replacement, any given member of the population can occur in the sample more than once. In a random sample without replacement, any given member of the population can be in the sample at most once. A random sample without replacement in which every subset of n of the N units in the population is equally likely is also called a simple random sample. The term random sample with replacement denotes a random sample drawn in such a way that every n-tuple of units in the population is equally likely. See also probability sample.

random triangulation

a triangulation of a surface or other manifold chosen by a random process. In a particularly interesting case, one considers all triangulations with a fixed, large number of triangles to be equally likely.

random variable

a random variable is an assignment of numbers to possible outcomes of a random experiment. For example, consider tossing three coins. The number of heads showing when the coins land is a random variable it assigns the number 0 to the outcome {T, T, T}, the number 1 to the outcome {T, T, H}, the number to the outcome {T, H, H}, and the number 3 to the outcome {H, H, H}.

randomised controlled experiment

an experiment in which chance is deliberately introduced in assigning subjects to the treatment and control groups. For example, we could write an identifying number for each subject on a slip of paper, stir up the slips of paper, and draw slips without replacement until we have drawn half of them. The subjects identified on the slips drawn could then be assigned to treatment, and the rest to control. Randomising the assignment tends to decrease confounding of the treatment effect with other factors, by making the treatment and control groups roughly comparable in all respects but the treatment.

■ range

the range of a set of numbers is the largest value in the set minus the smallest value in the set. Note that as a statistical term, the range is a single number, not a range of numbers.

■ rate

a ratio comparing two different units. For example, kilometres per hour and heartbeats per minute are rates.

■ ratio

a comparison expressed as indicated division. For example, there is a ratio of three boys to two girls in our class (3/2, 3:2).

■ rational map

a map from some field such as the real or complex numbers (plus infinity) to itself given by a rational polynomial function.

■ rational number

a number that can be expressed as a fraction or a ratio of two numbers where the denominator does not equal zero.

e.g., 5, ² /3, 0.4, 0.23.

■ ray

a line that has one end-point, but continues infinitely in the other direction.

■ real axis

the x-axis of an Argand diagram.

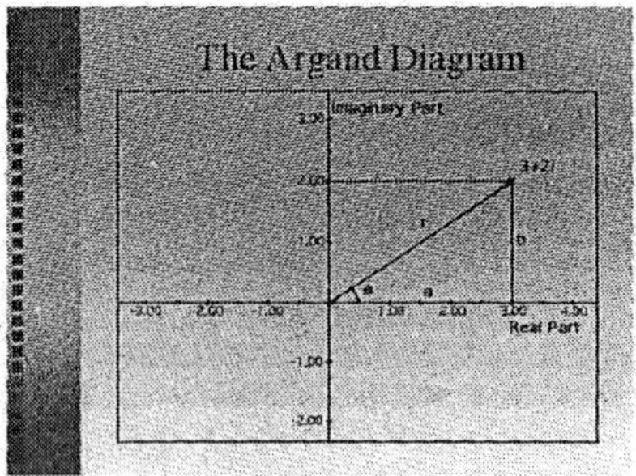

■ real number

1.any number, positive or negative, that is not imaginary. 4/3, 2.5, 0, and -8 are all examples of real numbers. 2. the union of the set of rational numbers and irrational numbers.

■ **real part**

the real number x is called the real part of the complex number x+iy where x and y are real and i=sqrt(-1).

■ **real variable**

a variable whose value ranges over the real numbers.

■ **reciprocal**

the reciprocal of a number, a, is 1/a, (a cannot be zero).

■ **reciprocal process**

a stochastic process whose state between any two times A and B depends only on the states at times A and B and not on any events before or after.

■ **recourse problem**

a two-stage stochastic optimisation problem in which one provides an initial choice, then a random event is revealed, and then one chooses a recourse. For example, an investor may select a stock to purchase, then observe the stock price for a period of time, and then sell part or all of the stock.

■ **rectangle**

a quadrilateral that has four 90 degree angles. Squares are a type of rectangle.

■ **rectangular array**

an organized arrangement of square units (tiles).

■ **rectangular coordinates**

same as Cartesian Coordinates.

■ **rectangular solid**

a solid figure consisting of 6 rectangular faces.

■ **recurrence relations**

in discrete mathematics, a value in a series is derived by applying a formula to the previous value.

■ **recursive function**

in discrete mathematics, a series of numbers in which values are derived by applying a formula to the previous value.

■ **recursive sequence**
in discrete mathematics, a series of numbers in which values are derived by applying a formula to the previous value.

■ **referent**
a point of reference used to compare in estimation.

■ **reflect**
in geometry, a transformation, also called a flip, which produces a mirror image of a geometric figure.

■ **reflection**
the reflection through a line in the plane or a plane in space is the transformation that takes each point in the plane to its mirror image with respect to the line or its mirror image with respect to the plane in space. It produces a mirror image of a geometric figure.

■ **reflex angle**
an angle with a degree measure between 180 and 360.

■ **reflexive property**
x = x for all x. Every number equals itself.

■ **regression fallacy**
the regression fallacy is to attribute the regression effect to an external cause.

■ **regression toward the mean, regression effect**
suppose one measures two variables for each member of a group of individuals, and that the correlation coefficient of the variables is positive (negative). If the value of the first variable for that individual is above average, the value of the second variable for that individual is likely to be above (below) average, but by fewer standard deviations than the first variable is. That is, the second observation is likely to be closer to the mean in standard units. For example, suppose one measures the heights of fathers and sons. Each individual is a (father, son) pair; the two variables measured are the height of the father and the height of the son. These two variables will tend to have a positive

correlation coefficient fathers who are taller than average tend to have sons who are taller than average. Consider a (father, son) pair chosen at random from this group. Suppose the father's height is 3SD above the average of all the fathers' heights. (The SD is the standard deviation of the fathers' heights.) Then the son's height is also likely to be above the average of the sons' heights, but by fewer than 3SD (here the SD is the standard deviation of the sons' heights).

■ **regression, linear regression**

linear regression fits a line to a scatterplot in such a way as to minimize the sum of the squares of the residuals. The resulting regression line, together with the standard deviations of the two variables or their correlation coefficient, can be a reasonable summary of a scatterplot if the scatterplot is roughly football-shaped. In other cases, it is a poor summary. If we are regressing the variable Y on the variable X, and if Y is plotted on the vertical axis and X is plotted on the horizontal axis, the regression line passes through the point of averages, and has slope equal to the correlation coefficient times the SD of Y divided by the SD of X. This page shows a scatterplot, with a button to plot the regression line.

■ **regular polygon**

a polygon in which all the angles are equal and all of the sides are equal.

■ **regular polyhedron**

a polyhedron whose faces are congruent, regular polygons.

■ **relation**

a set of ordered pairs.

■ **remainder**

what is left over after a number is divided into another number. For instance, in 12 divided by 5, the remainder is two.

rep digit
an integer all of whose digits are the same.

repeating decimal
a decimal that can be written using a horizontal bar to show the repeating digits.

representation theory
the study of linear representations of groups, Lie groups and Lie algebras.

representation/linear representation
a realisation of a group, Lie group, or Lie algebra by matrices or linear transformations. More technically, a homomorphism from a group to a group of matrices.

repunit
an integer consisting only of 1's.

residual
the difference between a datum and the value predicted for it by a model. In linear regression of a variable plotted on the vertical axis onto a variable plotted on the horizontal axis, a residual is the "vertical" distance from a datum to the line. Residuals can be positive (if the datum is above the line) or negative (if the datum is below the line). Plots of residuals can reveal computational errors in line r regression, as well as c nditions under which linear re gression is inappropriate, such as nonlinearity a d heteroscedasticity. If linear regression is performed properly, the sum of the res duals from the regressio line must be zero; otherwise, there is a computational error somewhere.

residual plot
a residual plot for a regression is a plot of the residuals from the regression against the explanatory variable.

resistant
a statistic is said to be resistant if corrupting a datum cannot change the statistic much. The mean is not resistant; the median is. See also breakdown point.

Reuleaux polytope
a convex body in the plane or in higher dimensions which, like a Reuleaux triangle, consists of pieces of round spheres, each centred at one of the corners of the convex body.

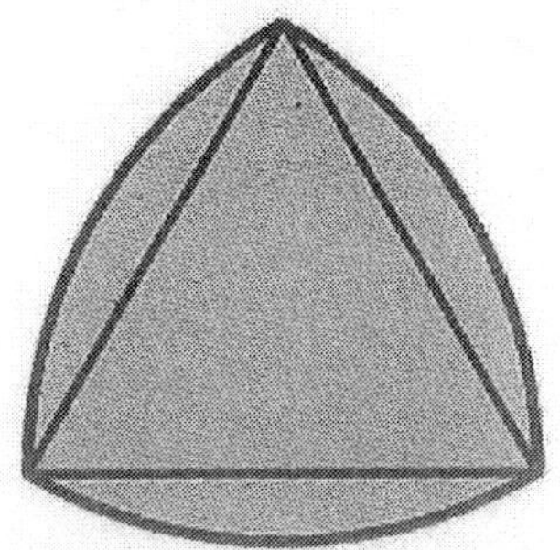

Reuleaux triangle
a three-cornered curve consisting of three equal arcs of circles, each centreed at the opposite corner.

rhombus
a quadrilateral that has equal sides but oblique angles, like a squashed square.

Riemann matrix
in the cohomology of a Riemann surface, the integral cohomology has a 2g x 2g generating matrix. After a standard basis manipulation using a natural inner product on the cohomology and its structure as a complex vector space, one obtains a g x g matrix called the Riemann matrix. It determines the Riemann surface and is unique up to transformations coming from homeomorphisms of the surface.

Riemann sphere
a topological sphere consisting of the complex plane and the point at infinity; an example of a Riemann surface.

Riemann surface/ complex curve
a surface with a conformal structure; a complex manifold with one complex dimension.

Riemannian geometry
the study of curvature and other properties of Riemannian metrics on manifolds. Sometimes also called differential geometry.

Riemannian metric
a metric on a manifold which is locally like ordinary distance in Euclidean space.

Here "locally" is not meant in the usual sense that every point has a region around it that is identical to a region in Euclidean space; rather, a Riemannian metric agrees with Euclidean distance to first order in the sense of calculus.

Riemann-Schottky problem

the problem of determining whether a given complex matrix is the Riemann matrix of some Riemann surface.

right angle

an angle formed by two perpendicular lines; a 90o angle.

right circular cone

a cone whose base is a circle located so that the line connecting the vertex to the center of the circle is perpendicular to the plane containing the circle.

right circular cylinder

a cylinder whose bases are circles and whose axis is perpendicular to its bases.

right triangle

a triangle that contains a right angle.

rigid motion

a transformation of the plane or space, which preserves distance and angles.

rms error of regression

the rms error of regression is the rms of the vertical residuals from the regression line. For regressing Y on X, the rms error of regression is equal to $(1 - r^2)^{1/2} \times SDY$, where r is the correlation coefficient between X and Y and SDY is the standard deviation of the values of Y.

root

the root of an equation is the same as the solution to the equation.

root extraction

finding a number that can be used as a factor a given number of times to produce the original number; for example, the fifth root of 32 = 2 because 2 x 2 x 2 x 2 x 2 = 32).

root of unity

a solution of the equation $x^n=1$, where n is a positive integer.

root-mean-square (rms)

the RMS of a list is the square-root of the mean of the squares of the elements in the list. It is a measure of the average "size" of the elements of the list. To compute the RMS of a list, you square all the entries, average the numbers you get, and take the square-root of that average.

root-mean-square error (rmse)

the RMSE of an an estimator of a parameter is the square-root of the mean squared error (MSE) of the estimator. In symbols, if X is an estimator of the parameter t, then.

rMSE(X) = (E((X-t)2))½.

the RMSE of an estimator is a measure of the expected error of the estimator. The units of RMSE are the same as the units of the estimator. See also mean squared error.

rose

the graph in a polar coordinates of r= a sin nØ, or r= a cos nØ, where n is a positive integer. It consists of rose petal-shaped loops with the origin a point common to all of them. When n is odd there are n loops; when n is even there are 2n loops.

rotation

a rotation in the plane through an angle q and about a point P is a rigid motion T fixing P so that if Q is di tinct from P, then the angle between the lines PQ and PT(Q) is always Φ. A rotation through an angle Φ in space is a rigid motion T fixing the points of a line l so that it is a rotation through ? in the plane perpendicular to l through some point on l.

round

in topology, the terms circle and sphere refer to topological objects and not geometric ones, so that the surface of an egg shape is a sphere.

A round sphere, then, is a sphere with constant curvature; a sphere in the sense of geometry.

■ **rounding**

to give a close approximation of a number by dropping the least significant numbers. For example 15.88 can be rounded up to 15.9 (or 16) and 15.12 can be rounded down to 15.1 (or 15).

■ **round-off error**

the error accumulated during a calculation due to rounding intermediate results.

■ **ruled surface**

a surface formed by moving a straight line (called the generator).

■ **Russell Paradox(Bertrand Russell, 1901)**

a paradox of set theory which necessitated a more careful axiomatization of set theory in the 1920's and 1930's: Naively, some sets are members of themselves and some are not. For instance, the set of all apples is not itself an apple, but the set of all sets does seem to be a set. So consider the set X of all sets that are not members of themselves. We may ask, is X a member of itself? If it is then it cannot be, because of the way in which X itself was defined, but if it isn't then it must be, by the same reasoning. Contradiction. The Russell paradox is resolved in modern set theory by a foundation axiom or axiom of regularity, and by limiting the "size" of objects. we call sets. For example, the "set of all sets" is considered not to be a set but a proper class.

■ **Russell, Bertrand(born 1872)**

english philosopher and logician. Wrote Principia Mathematica (1913) with Alfred North Whitehead, an attempt to reduce all of mathematics to symbolic logic. This was perhaps the most important effort in the logicist program. Russell introduced type theory into the theory of sets and classes in an effort to avoid the kind of antinomy in the foundations of mathematics as was exemplified in the so-called Russell paradox. The goal of Russell's system was to show that any true mathematical proposition can be established by logic alone, a goal which was severely compromised by Kurt Gödel's proof of the Gödel Incompleteness Theorem in 1931.

■ rusty compass

a pair of compasses that are fixed open in a given position.

■ saddle function

a function F(x,y) of two vectors x and y (which typically lie in different vector spaces) is a saddle function if it is concave up in x and concave down in y.

■ sample

a part of the total population. Used in statistics to make predictions about the characteristics of the entire group.

■ sample mean

the arithmetic mean of a random sample from a population. It is a statistic commonly used to estimate the population mean. Suppose there are n data, {x1, x2, ... , xn}. The sample mean is (x1 + x2 + ... + xn)/n. The expected value of the sample mean is the population mean. For sampling with replacement, the SE of the sample mean is the population standard deviation, divided by th square-root of the sample size. For sampling without replacement, the SE of the sample mean is the finite-population correction ((N-n)/(N-1))½ times the SE of the sample mean for sampling with replacement, with N the size of the population and n the size of the sample.

■ sample percentage

the percentage of a random sample with a certain property, such as the percentage of voters registered as Democrats in a simple random sample of voters. The sample mean is a statistic commonly used to estimate the population percentage. The expected value of the sample percentage from a simple random sample or a random sample with replacement is the population percentage. The SE of the sample percentage for sampling with replacement is $(p(1-p)/n)^{1/2}$, where p is the population percentage and n is the sample size. The SE of the sample percentage for sampling without replacement is the finite-population correction $((N-n)/(N-1))^{1/2}$ times the SE of the sample percentage for sampling with replacement, with N the size of the population and n the size of the sample. The SE of the sample percentage is often estimated by the bootstrap.

■ sample size

the number of elements in a sample from a population.

■ sample standard deviation, s

the sample standard deviation S is an estimator of the standard deviation of a population based on a random sample from the population. The sample standard deviation is a statistic that measures how "spread out" the sample is around the sample mean. It is quite similar to the standard deviation of the sample, but instead of averaging the squared deviations (to get the rms of the deviations of the data from the sample mean) it divides the sum of the squared deviations by (number of data - 1) before taking the square-root. Suppose there are n data, $\{x1, x2, \ldots, xn\}$, with mean $M = (x1 + x2 + \ldots + xn)/n$. Then.

$s = (((x1 - M)2 + (x2 - M)2 + \ldots + (xn - M)2)/(n-1))^{1/2}$

the square of the sample standard deviation, S2 (the sample variance) is an unbiased estimator of the square of the SD of the population (the variance of the population).

■ sample sum

the sum of a random sample from a population. The expected value of the sample sum is the sample size times the population mean. For sampling with replacement, the SE of the sample sum is the population standard deviation, times the square-root of the sample size. For sampling without replacement, the SE of the sample sum is the finite-population correction ((N-n)/(N-1))½ times the SE of the sample sum for sampling with replacement, with N the size of the population and n the size of the sample.

■ sample survey

a survey based on the responses of a sample of individuals, rather than the entire population.

■ sample variance

the sample variance is the square of the sample standard deviation S. It is an unbiased estimator of the square of the population standard deviation, which is also called the variance of the population.

■ sampling distribution

the sampling distribution of an estimator is the probability distribution of the estimator when it is applied to random samples. The tool on this page allows you to explore empirically the sampling distribution of the sample mean and the sample percentage of random draws with or without replacement draws from a box of numbered tickets.

■ sampling error

in estimating from a random sample, the difference between the estimator and the parameter can be written as the sum of two components bias and sampling error. The bias is the average error of the estimator over all possible samples. The bias is not random. Sampling error is the component of error that varies from sample to sample. The sampling error is random it comes from "the luck of the draw" in which units happen to be in the sample. It is the chance variation of the estimator. The average of the sampling error over all possible samples (the expected value of the sampling error) is zero. The standard error of the estimator is a measure of the typical size of the sampling error.

■ sampling unit

a sample from a population can be drawn one unit at a time, or more than one unit at a time (one can sample clusters of units). The fundamental unit of the sample is called the sampling unit. It need not be a unit of the population.

■ scalar

a real number. Also a quantity that has magnitude but no direction, such as mass and density.

■ scalar curvature

part of the curvature of a Riemannian manifold at a given point p, or the corresponding real-valued function of p. It can be defined as the limiting discrepancy between the volume of a small metric ball at p and the volume in Euclidean space of a ball the same size, times a suitable proportionality constant.

■ scalar curvature equation

a partial differential equation arising in differential geometry. Given a function f on a manifold with a conformal structure, a solution to the equation produces a Riemannian metric compatible with the conformal sturcture and with scalar curvature f.

■ scalar matrix

a diagonal matrix whose elements are equal. Multiplication by such a matrix is equivalent to multiplication by a scalar (a real number, a quantity with magnitude but no direction). The unit matrix is a special case.

■ scalene triangle

a triangle with three unequal sides.

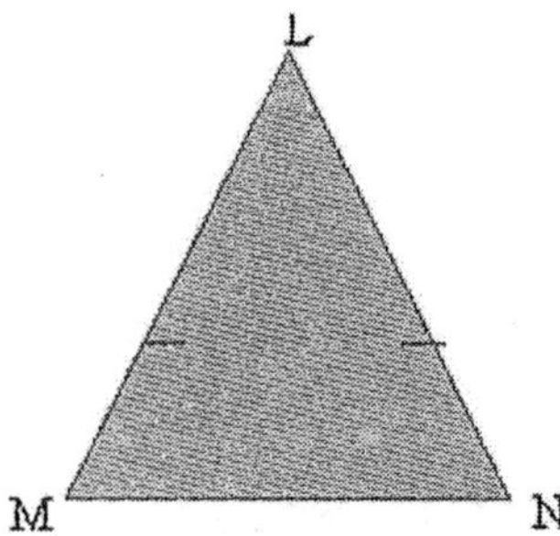

scatterplot

a scatterplot is a way to visualise bivariate data. A scatterplot is a plot of pairs of measurements on a collection of "individuals" (which need not be people). For example, suppose we record the heights and weights of a group of 100 people. The scatterplot of those data would be 100 points. Each point represents one person's height and weight. In a scatterplot of weight against height, the x-coordinate of each point would be height of one person, the y-coordinate of that point would be the weight of the same person. In a scatterplot of height against weight, the x-coordinates would be the weights and the y-coordinates would be the heights.

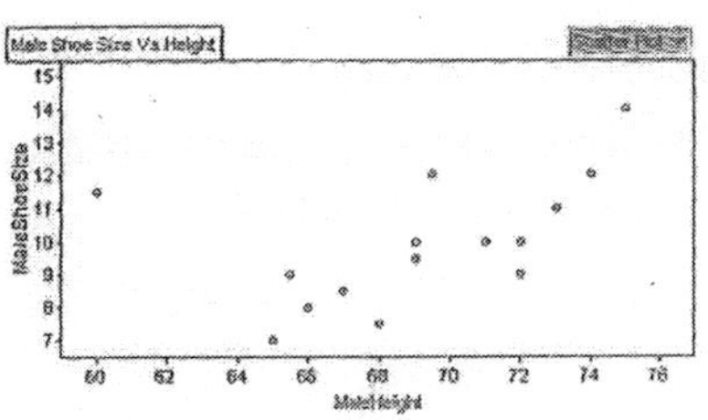

scenario analysis

a method in stochastic optimisation in which one finds the optimum solution in a sample of cases (scenarios) and then iteratively reconciles the separate solutions to obtain a common solution to the stochastic programming problem.

Schroeder-Bernstein Theorem

if there exists an injection from a set X into a set Y, and also an injection from Y into X, then there exists a bijection from X to Y, and hence X and Y have the same cardinality.

scientific notation

a shorthand way of writing very large or very small numbers. A number expressed in scientific notation is expressed as a decimal number between 1 and 10 multiplied by a power of 10 (e.g., 7000 = 7 x 103 or 0.0000019 = 1.9 x 10-6).

sd line

for a scatterplot, a line that goes through the point of

averages, with slope equal to the ratio of the standard deviations of the two plotted variables. If the variable plotted on the horizontal axis is called X and the variable plotted on the vertical axis is called Y, the slope of the SD line is the SD of Y, divided by the SD of X.

■ **secant**

a straight lien that meets a curve in two or more points.

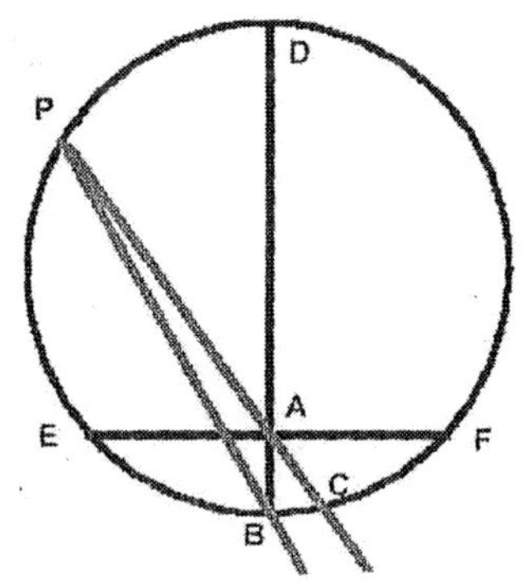

■ **second**

the unit of measure of an angle that is 1/60 of a minute.

■ **sector**

a region bounded by two radii of a circle and the arc whose endpoints lie on those radii.

■ **secular trend**

a linear association (trend) with time.

■ **segment**

the union of a point, A, and a point, B, and all the points between them.

■ **selection bias**

a systematic tendency for a sampling procedure to include and/or exclude units of a certain type. For example, in a quota sample, unconscious prejudices or predilections on the part of the interviewer can result in selection bias. Selection bias is a potential problem whenever a human has latitude in selecting individual units for the sample; it tends to be eliminated by probability sampling schemes in which the interviewer is told exactly whom to contact (with no room for individual choice).

■ **self-selection**

self-selection occurs when individuals decide for themselves whether they are in the control group or the treatment group. Self-selection is quite common in studies of human behavior. For example, studies of the effect of smoking on human health involve self-selection individuals choose for themselves whether or not to smoke. Self-selection precludes an experiment; it results in an observational study. When there is self-selection, one must be wary of possible confounding from factors that influence individuals' decisions to belong to the treatment group.

semi-infinite form

a mathematical object suggested by the Dirac sea of electrons in quantum mechanics. More technically, an algebraic structure, often used as a vector space or a representation, of infinitely many anti-commuting variables divided into two camps, positive and negative. The typical vector is a product of all but finitely many of the negative variables- and finitely many of the posi ive ones.

semi-magic square

a square array of n numbers such that sum of the n numbers in any row or column is a constant (known as the magic sum).

sequence

an ordered set of numbers derived according to a rule, each member being determined either directly or from the preceding terms.

series

the sum of a finite or infinite sequence.

set

a set is a collection of things, without regard to their order.

shallow water

any of several mathematical models of waves in water that add terms caused by the finite depth of the water.

shock wave

mathematically, a wave front in a non-linear PDE with a discontinuous change in pressure or some other parameter. Physically, a similar wave with a very abrupt change in pressure. Typically a shock wave can exist because the speed of communication behind it is greater than the speed of communication in front of it.

■ **Siegel upper half-space**

the set of n x n matrices for some n with positive-definite imaginary part. It is considered together with a group action taking Z to (AZ+B)/(CZ+D), where A,B,C, and D are certain integral matrices.

■ **sigma (s, s)**

represents summation.

■ **significance, significance level, statistical significance**

the significance level of an hypothesis test is the chance that the test erroneously rejects the null hypothesis when the null hypothesis is true.

■ **significant digit**

any digit that is obtained by actual measurement and thus is not simply a placeholder used to position the decimal point.

■ **similar figures**

two geometric figures are similar if their sides are in proportion and all their angles are the same.

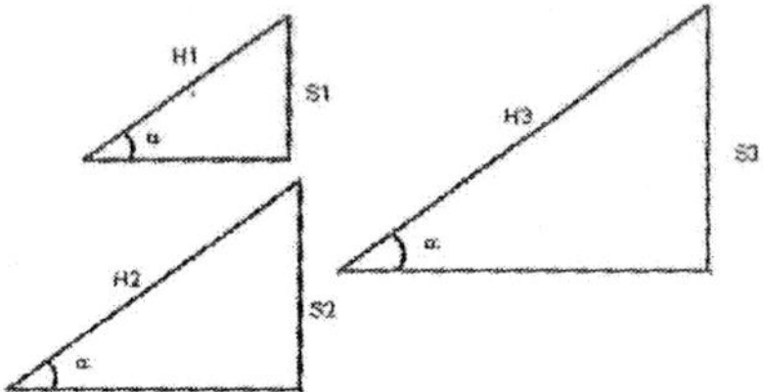

■ **similarity**

in geometry, two shapes R and S are similar if there is a dilation D (see the definition of dilation) that takes S to a shape congruent to R. It follows that R and S are similar if they are congruent after one of them is expanded or shrunk.

■ **simple exclusion process**

a statistical model and cellular automaton consisting of a collection of particles on a grid, usually filling a finite fraction of the grid. Each particle has a Poisson timer which triggers it to attempt to move to an adjacent location, and it will do so unless another particle is already there. In the second case, it does not move.

simple random sample

a simple random sample of n units from a population is a random sample drawn by a procedure that is equally likely to give every collection of n units from the population; that is, the probability that the sample will consist of any given subset of n of the N units in the population is 1/NCn. Simple random sampling is sampling at random without replacement (without replacing the units between draws). A simple random sample of size n from a population of N >= n units can be constructed by assigning a random number between zero and one to each unit in the population, then taking those units that were assigned the n largest random numbers to be the sample.

simple recourse

a recourse problem in which the optimal recourse is simply and explicitly determined by the information revealed after the initial choice.

simplex

the n-dimensional generalisation of the triangle and the tetrahedron; a polytope in n dimensions with n+1 vertices.

simplicial

made up of simplices. For example, a simplicial polytope has simplices as faces and a simplicial complex is a collection of simplices pasted together in any reasonable vertex-to-vertex and edge-to-edge arrangement.

simpson's paradox

what is true for the parts is not necessarily true for the whole. See also confounding.

simulataneous equations

a group of equations that are all true at the same time.

simultaneous linear equations

pair of equations of the first degree upon which two different conditions are put on the same two variables at the same time (e.g., find two numbers whose sum is 7 and whose difference is 1. x + y= 7 and x - y= 1 The solution to the set is the single pair x=4, y=3).

■ sine
Sin(Φ) is the y-coordinate of the point on the unit circle so that the ray connecting the point with the origin makes an angle of Φ with the positive x-axis. When Φ is an angle of a right triangle, then sin(Φ) is the ratio of the opposite side with the hypotenuse.

■ sine-Gordon equation/ sinh-Gordon equation
the partial differential equation Nabla u + sin(u) = 0, or any of several variants, such as the one obtained by replacing sine by hyperbolic sine. All such equations are non-linear variants of the Klein-Gordon equation.

■ singular perturbation
the addition of an extra small term in an equation, especially a differential equation, that qualitatively changes the nature of its solutions.

■ singular vector
a vector in a Verma module which is annihilated by all raising operators; a vector which generates a Verma submodule.

■ skein theory
an inductive definition of an invariant of knots and links which postulates a linear relation between the invariant of a given link and the invariant of the same link with a crossing switched or otherwise simplified. Sometimes skein theories also involve tangled graphs.

■ skeleton division
a long division in which most or all of the digits have been replaced by asterisks to form a cryptarithm.

■ skew lines
two lines in three-dimensional space which do not lie in the same plane (and do not intersect).

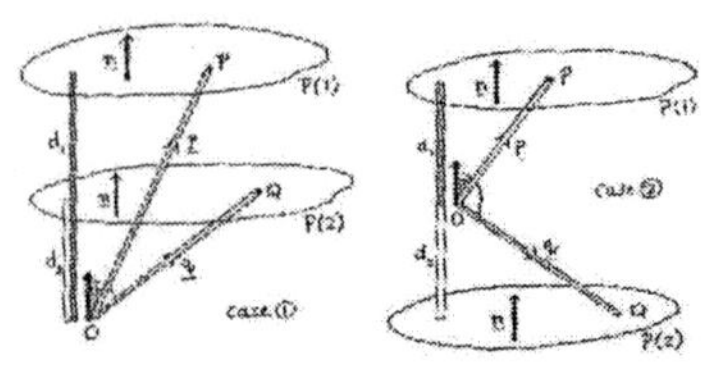

skewed distribution

a distribution that is not symmetrical.

skinning map

an iterative map on conformal structures on a surface that appears in Thurston's proof of the Geometrisation Theorem. In the proof, the hyperbolic structure of a cut-open manifold must be varied until it fits with itself when the manifold is glued back together. Such a hyperbolic structure is determined by a conformal structure at the wounds of the cut (ends). Given one such structure A, the skinning map produces a new one B chosen to match A; structure B also comes closer to matching itself.

skip counting

counting by equal intervals.

slide rule

a calculating device consisting of two sliding logarithmic scales.

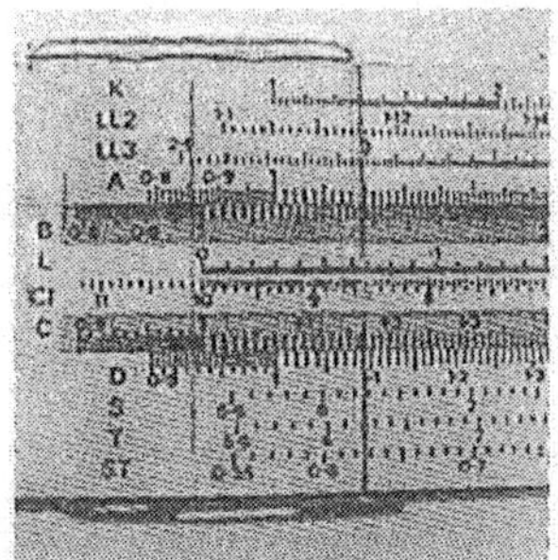

slides

in geometry, a transformation where a figure moves in a given direction.

slope

the slope of a line is the change in the vertical coordinates/the change in the horizontal coordinates of any two points on the line.

smooth

1. Infinitely differentiable; possessing infinitely many derivatives. For example, sin(x) is a smooth function, while |x|^3 is not. More complicated mathematical objects such as manifolds are called smooth if they are defined or described by smooth functions.

2. Continuously differentiable; possessing a continu-

ous tangent or derivative.
solid
a three dimensional object that completely encloses a volume of space.

solid geometry
the area of geometry that involves 3d objects.

solid of revolution
a solid formed by rotation a plane figure about an axis in three-space.

solidus
the slanted line in a fraction such as a/b dividing the numerator from the denominator.

soliton
a solution to a partial differential equation which is localised in some directions but not localised in time, and which does not change its shape. An example of a traveling wave.

space
1. A general term referring to many kinds of mathematical objects. When an object is called a space, there is a Platonist connotation that some kind of being could exist in it and examine its properties as if it were a jail or the entire universe.
2. More specifically, 3-dimensional Euclidean space.

space geometry
the study of three-dimensional figures. Sometimes called solid geometry.

special numbers
numbers that are near "special" values that are easy to compute mentally. e.g., ½, 1, 50.

spectrum
1. in quantum physics, the set of allowed energy levels of a particle or system. It is directly related to bright or dark lines in a spectrum of light produced by a prism.
2. in mathematics, the set of eigenvalues of a linear transformation. By historical coincidence, it is equivalent to the notion of a spectrum in quantum mechanics.

sphere
1. the locus of points in three-space that are a fixed

distance from a given point (called the centre).
2. a three-dimensional object in which all points are equidistant from the centre (a ball shape).

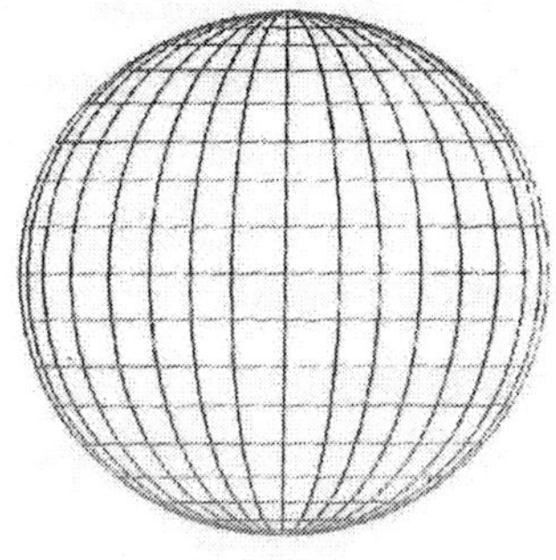

The surface of a sphere.

■ spherical trigonometry
the branch of mathematics dealing with measurements on the sphere.

■ spread, measure of
see also inter-quartile range, range, and standard deviation.

■ square
a quadrilateral with 4 equal sides and 4 right angles.

■ square free
an integer is said to be square free if it is not divisible by a perfect square, n2, for n>1.

■ square number
a number of the form n^2.

■ square root
the square roots of n are all the numbers m so that m2= n. The square roots of 16 are 4 and -4. The square roots of -16 are 4i and -4i.

■ square-root law
the Square-Root Law says that the standard error (SE) of the sample sum of n random draws with replacement from a box of tickets with numbers on them is.

sE(sample sum) = $n^{1/2}$×SD(box),

and the standard error of the sample mean of n random draws with replacement from a box of tickets is.

sE(sample mean) = $n^{-1/2}$×SD(box),

where SD(box) is the standard deviation of the list of the numbers on all the tickets in the box (including repeated values).

■ standard deviation (sd)
the standard deviation of a set of numbers is the rms of

the set of deviations between each element of the set and the mean of the set. See also sample standard deviation.

■ standard error (se)

the Standard Error of a random variable is a measure of how far it is likely to be from its expected value; that is, its scatter in repeated experiments. The SE of a random variable X is defined to be. sE(X) = [E((X - E(X))2)] ½.

that is, the standard error is the square-root of the expected squared difference between the random variable and its expected value. The SE of a random variable is analogous to the SD of a list.

■ standard normal curve

see normal curve.

■ standard units

a variable (a set of data) is said to be in standard units if its mean is zero and its standard deviation is one. You transform a set of data into standard units by subtracting the mean from each element of the list, and dividing the results by the standard deviation. A random variable is said to be in standard units if its expected value is zero and its standard error is one. You transform a random variable to standard units by subtracting its expected value then dividing by its standard error.

■ standardise

to transform into standard units.

■ stationary point

point at which the derivative of a function is zero. Includes maximum and minimum turning points, but not all stationary points are turning points.

■ statistic

a number that can be computed from data, involving no unknown parameters. As a function of a random sample, a statistic is a random variable. Statistics are used to estimate parameters, and to test hypotheses.

statistical estimation
the problem of estimating a probability distribution given a set of random samples from the distribution. For example, estimating the distribution of income of people in a city given the results of selective poll.

statistical mechanics
the study of statistical and thermal properties of physical materials and their idealised mathematical models.

stem-and-leaf plot
a table utilizing digit(s) of a number as stems and the other digit(s) as leaves. For example, 5 | 7, 8 shows 57 and 58.

stochastic differential equation
a differential equation whose coefficients change randomly in the time parameter of the equation. Typically it is solved by producing a range of solutions with a probability distribution.

stochastic optimisation
the study of maximising or minimising (optimising) desirable quantities (such as time or money) when the constraints on the allowed choices are chosen randomly. The function one wishes to maximise may also depend on random parameters. For example, the problem of choosing raw materials in manufacturing when their prices might change unpredictably.

stochastic process
a dynamical system with random fluctations at each iteration or influenced by random noise. A random variable which, at each stage in time, depends on its previous values and on further random choices. For example, the price of a stock is often modelled as a stochastic process.

stochastic programming
the theory of linear programming in the context of stochastic optimisation; the

problem of maximising a randomly chosen linear function on a randomly chosen polytope.

■ **Stone-Weierstrass Theorem**

if X is a compact space and C(X) denotes the space of all continuous functions on X, and A is an algebra of functions in C(X) which separates the points of C(X) and which contains a constant function f not identically zero, then A is dense in C(X).

■ **straight angle**

an angle with a measure of 180 degrees, appears like a line.

■ **stratifiable**

a topological space is stratifiable if it can be decomposed into shells which are similar to the shells that are a fixed distance from a point in a metric space.

■ **stratified sample**

in a stratified sample, subsets of sampling units are selected separately from different strata, rather than from the frame as a whole.

■ **stratified sampling**

the act of drawing a stratified sample.

■ **stratum**

in random sampling, sometimes the sample is drawn separately from different disjoint subsets of the population. Each such subset is called a stratum. (The plural of stratum is strata.) Samples drawn in such a way are called stratified samples. Estimators based on stratified random samples can have smaller sampling errors than estimators computed from simple random samples of the same size, if the average variability of the variable of interest within strata is smaller than it is across the entire population; that is, if stratum membership is associated with the variable. For example, to determine average home prices in the U.S., it would be advantageous to stratify on geography, because average home prices vary enormously with location. We might divide the country into states, then divide each state into urban, suburban, and rural areas; then draw random samples separately from each such division.

string theory

a physical theory in which one-dimensional loops travel through space and al o merge and lyse as time elaps es. This is in contrast to ordinary quantum field the ory, which predicts point - particles that emit and absorb each other. String theory is a candidate for a Theory of Everything.

studentised score

the observed value of a statistic, minus the expected value of the statistic, divided by the estimated standard error of the statistic.

student's T curve

student's t curve is a family of curves indexed by a parameter called the degrees of freedom, which can take the values 1, 2, ... Student's t curve is used to approximate some probability histograms. Consider a population of numbers that are nearly normally distributed and have population mean is μ. Consider drawing a random sample of size n with replacement from the population, and computing the sample mean M and the sample standard deviation S. Define the random variable $t = (M - \mu)/(S/n^{1/2})$. If the sample size n is large, the probability histogram of T can be approximated accurately by the normal curve. However, for small and intermediate values of n, Student's t curve with n - 1 degrees of freedom gives a better approximation. That is, $P(a < T < b)$ is approximately the area under Student's T curve with n - 1 degrees of freedom, from a to b. Student's t curve can be used to test hypotheses about the population mean and construct confidence intervals for the population mean, when the population distribution is known to be nearly normally distributed. This page contains a tool that shows Student's t curve and lets you find the area under parts of the curve.

SU(2)

the 3-dimensional Lie group of 2 x 2 unitary matrices; the most common Lie group in mathematics and physics after the circle.

subject, experimental subject

a member of the control group or the treatment group.

subset

a subset of a given set is a collection of things that belong to the original set. Every element of the subset must belong to the original set, but not every element of the original set need be in a subset (otherwise, a subset would always be identical to the set it came from).

subtend

to lie opposite and mark out the limits of an angle.

subtraction

the inverse operation of addition. In the notation a - b = c, the terms a, b, and c are called the minuend, subtrahend and difference, respectively.

subtrahend

the number to be subtracted. e.g., 8 - 3 = 5, 3 is the subtrahend.

sum

the result after addition. Four is the sum of two plus two.

super

a prefix that means that a mathematical structure is analogous to supersymmetry. Typically these structures have commuting and anticommmuting variables that are placed on an equal footing. For example, a supermanifold has anticommuting supercoordinates along with the usual commuting coordinates of its coordinate charts.

superset

a larger set containing smaller sets.

■ **supersymmetry**
a theory in physics that postulates a counterintuitive symmetric relationship between fermions, which are particles such as electrons that obey the Pauli exclusion principle, and bosons, which are particles such as photons that enjoy being in the same state as each other.

■ **supplementary**
two angles are supplementary if their sum is 180 degrees.

■ **supplementary angle**
either of two angles that produce 180 degrees when added.

■ **survey**
see sample survey.

■ **symmedian**
reflection of a median of a triangle about the corresponding angle bisector.

■ **symmetric**
two points are symmetric with respect to a third point if the segments joining them to the third point are equal. Two points are symmetric with respect to a line if the line is the perpendicular bisector of the segment joining the points.

■ **symmetric distribution**
the probability distribution of a random variable X is symmetric if there is a number a such that the chance that X>=a+b is the same as the chance that X<=a-b for every value of b. A list of numbers has a symmetric distribution if there is a number a such that the fraction of numbers in the list that are greater than or equal to a+b is the same as the fraction of numbers in the list that are less than or equal to a-b, for every value of b. In either case, the histogram or the probability histogram will be symmetrical about a vertical line drawn at x=a.

■ **symmetrical**
a shape that can be folded in half so that the two parts match.

symmetry

1.a symmetry of a shape S in the plane or space is a rigid motion T that takes S onto itself (T(S) = S). For example, reflection through a diagonal and a rotation through a right angle about the center are both symmetries of the square.

2.a correspondence in size, form and arrangement of parts on opposite sides of a plane, line or point. For example, a figure that has line symmetry has two halves which coincide if folded along its line of symmetry.

synthetic representation

the geometric form as opposed to the algebraic representation of a figure.

system of linear equations

set of equations of the first degree (e.g., $x + y = 7$ and $x - y = 1$). A solution of a set of linear equations is a set of numbers a, b, c, ... so that when the variables are replaced by the numbers all the equations are satisfied. For example, in the equations above, $x = 4$ and $y = 3$ is a solution.

systematic error

an error that affects all the measurements similarly. For example, if a ruler is too short, everything measured with it will appear to be longer than it really is (ig noring random error). If you watch runs fast, every time in erval you measure with it wil appear to be longer than it really is (again, ignoring random error). Systematic errors do not tend to average out.

systematic random sample

a systematic sample starting at a random point in the listing of units in the of frame, instead of starting at the first unit. Systematic random sampling is better than systematic sampling, but typically not as good as simple random sampling.

systematic sample

a systematic sample from a frame of units is one drawn by listing the units and selecting every kth element of the list. For example, if there are N units in the frame, and we want a sample of size N/10, we would take every tenth unit the first unit, the eleventh unit, the 21st unit, etc. Systematic samples are not random samples, but they often behave essentially as if they were random, if the order in which the units appears in the list is haphazard. Systematic samples are a special case of cluster samples.

t test

an hypothesis test based on approximating the probability histogram of the test statistic by Student's t curve. t tests usually are used to test hypotheses about the mean of a population when the sample size is intermediate and the distribution of the population is known to be nearly normal.

tangent

a trigonometric function of an angle which is defined as the ratio of the lengths of the leg opposite to the leg adjacent to an angle in its right triangle. Also a line having one point in common with a curve.

tangled graph

a graph in 3-dimensional space; equivalently, a graph drawn in the plane so that when edges cross, one edge goes over the other.

tangram

a puzzle made by cutting a square into triangles, a square and a rhomboid which can be recombined in many different figures.

tau function

a form for the 2-point correlations of an integrable system that typically satisfies auxiliary differential equations and is also given by determinant formulas.

■ tautology

a sentence that is true because of its logical structure.

■ ten-frame

a 2 x 5 array of squares used to teach counting, number relationships and computation.

■ tensor

an object in linear algebra, generalising a vector, an inner product, and a linear transformation, with a multidimensional matrix.

■ term

a part of a sum in an algebraic expression.

■ terminating decimal

a decimal that contains a finite number of digits. For example, 0.408 is a terminating decimal.

■ terminating decimal

a fraction whose decimal representation contains a finite number of digits.

■ terms of a fraction

the elements of a fraction. For example, in 3/4 , 3 and 4 are both terms.

■ tessellation

a tiling, made up of the repeated use of geometric figures, to completely fill a plane without gaps or overlapping.

■ test statistic

a statistic used to test hypotheses. An hypothesis test can be constructed by deciding to reject the null hypothesis when the value of the test statistic is in some range or collection of ranges. To get a test with a specified significance level, the chance when the null hypothesis is true that the test statistic falls in the range where the hypothesis would be rejected must be at most the specified significance level. The Z statistic is a common test statistic.

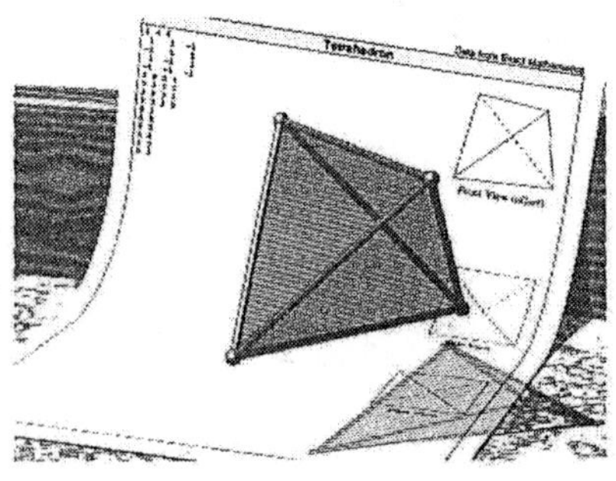

■ **tetromino**
a four-square polyomino.

■ **theorem**
a statement that has been proven.

■ **theoretical (mathematical)**
relating to the probability of a given event, using mathematical relationships (e.g., the chance of a red side coming up on the flip of a two-colored counter is one in two or 1/2).

■ **theory of everything**
a unified theory of all fundamental forces and interactions in nature; a Grand Unified Theory that also includes gravity or general relativity.

■ **thin position**
a representation of a knot or graph in Euclidean space, or some analogous structure, with the simplest possible horizontal slices. Between U-turns where the knot or link has a horizontal tangent (and between vertices of the graph), one counts the number of intersections with a horizontal plane. Thin position is achieved when the total of these counts is minimised.

■ **three-dimensional figure**
any figure that contains depth as well as length and width.

■ **threshold growth**
a simple cellular automaton in which cells may be alive or dead, a cell is born if sufficiently neighbours are alive, and a cell, once alive, is immortal.

■ **threshold vote**
a cellular automaton in which each cell holds one of several opinions, and at each iteration it is moved to change its mind if sufficiently many neighbours hold a different opinion. If more than one opinion in the neighbourhood meets the threshold, the cell retains its old opinion.

■ **Tietze extension theorem**
the theorem that if X is a closed subset of a metrisable space Y (or more generally a

normal space), a continuous, real-valued function on X can be extended to Y.

■ **Toeplitz matrix**
a matrix in which all the elements are the same along any diagonal that slopes from northwest to southeast.

■ **topological complexity**
a lower bound on computational complexity introduced by Smale that involves topology. A task with high computational complexity requires a computer to make many decisions, sometimes arbitrary decisions, to untangle the topology of the space of possible inputs or outputs or the space of possibilities for an intermediate quantity.

■ **topological quantum field theory**
a quantum field theory, such as the Chern-Simons field theory in three dimensions, whose integrals for a manifold produce topological invariants.

■ **topological space**
the basic object of topology; a formal model of the qualitative way in which something is connected to itself. The technical definition is that a topological space is a set X together with distinguished subsets called open sets or open neighbourhoods, such that the the empty set and X are open, the intersection of two open sets is open, and the union of an arbitrary collection of open sets is open. See also the definition of space.

■ **topology**
the study of how geometric objects are intrinsically connected to themselves. Since topologists are not concerned with the geometric measurements of objects, people often say that they study objects up to continuous deformation. But usually topologists consider spaces which have a topology (a qualitative shape or connectivity) but no predefined (quantitative) geometry. Knots and manifolds are typical examples of topological spaces.

■ **torus**

a geometric solid in the shape of a donut.

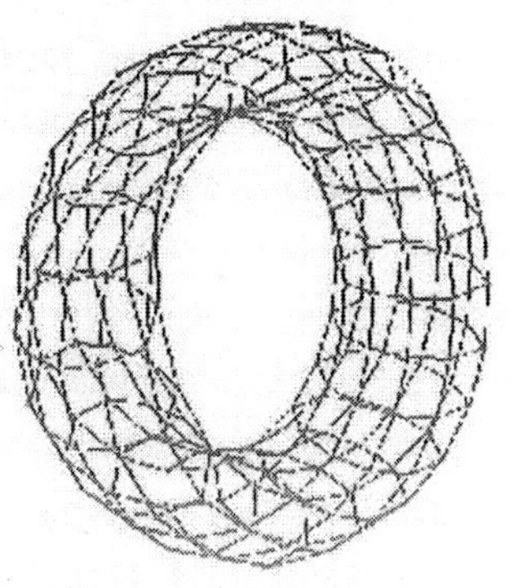

■ **trace**

the trace of a matrix is the sum of the terms along the principal diagonal.

■ **train track**

a graph drawn on a surface such that every vertex has degree three, and such that all three edges meeting at a vertex have a common tangent, two edges on one side and one on the other. Every lamination of a surface is approximately parallel to, or carried by, a train track.

■ **trajectory**

the path that a body makes as it moves through space.

■ **transcendental function**

a function which is not an algebraic function, i.e., a function whose action on its argument(s) cannot be represented by the arithmetic and algebraic operations: addition and subtraction, multiplication and division, raising to a power, or extraction of roots. The exponential function, the logarithmic f nction, and the trigonometric functions are examples of transcendental functions.

■ **transcendental number**

a real number that does not satisfy any algebraic equation with integral coefficients, such as $x^3 - 5x + 11 = 0$. All transcendental numbers are irrational and most irrational numbers (non-repeating, non-terminating decimals) are transcendental. Transcendental functions (such as exponential, sine and cosine functions) can burst into chaos under certain circumstances.

transformation

transformations turn lists into other lists, or variables into other variables. For example, to transform a list of temperatures in degrees Celsius into the corresponding list of temperatures in degrees Fahrenheit, you multiply each element by 9/5, and add 32 to each product. This is an example of an affine transformation multiply by something and add something (y = ax + b is the general affine transformation of x; it's the familiar equation of a straight line). In a linear transformation, you only multiply by something (y = ax). Affine transformations are used to put variables in standard units. In that case, you subtract the mean and divide the results by the SD. This is equivalent to multiplying by the reciprocal of the SD and adding the negative of the mean, divided by the SD, so it is an affine transformation. Affine transformations with positive multiplicative constants have a simple effect on the mean, median, mode, quartiles, and other percentiles the new value of any of these is the old one, transformed using exactly the same formula. When the multiplicative constant is negative, the mean, median, mode, are still transformed by the same rule, but quartiles and percentiles are reversed the qth quantile of the transformed distribution is the transformed value of the 1-qth quantile of the original distribution (ignoring the effect of data spacing). The effect of an affine transformation on the SD, range, and IQR, is to make the new value the old value times the absolute value of the number you multiplied the first list by what you added does not affect them.

transitive property

the property that states that if a = b, and b = c, then a = c.

translate

a rigid motion that preserves exactly the size, shape, and orientation of a figure.

■ **translation**

a rigid motion of the plane or space of the form X goes to X + V for a fixed vector V.

■ **transport**

any mechanism in physics by which particles or regions of fluid move around, or a mathematical model of su h a mechanism such as a PDE.

■ **transport coefficient**

any of various parameters in a mathematical model of transport of particles. More technically, a coefficient of a transport term in a PDE.

■ **transversal**

in geometry, given two or more lines in the plane a transversal is a line distinct from the original lines and intersects each of the given lines in a single point.

■ **trapezium**

a quadrilateral in which no sides are parallel.

■ **trapezoid**

a four-sided figure with one pair of opposite sides parallel.

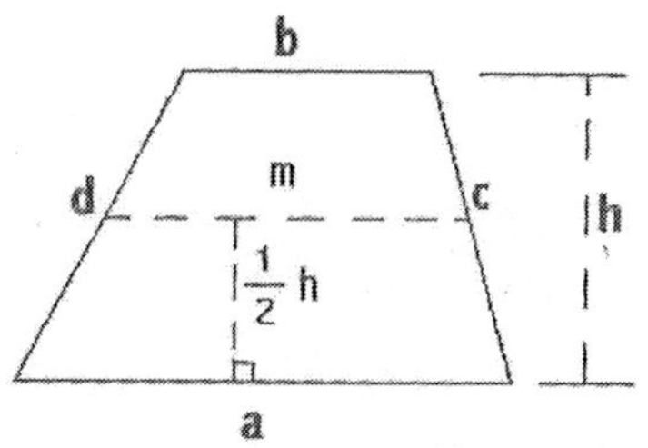

■ **traveling wave**

any wave, such as a soliton, which does not change shape as it moves.

■ **treatment**

the substance or procedure studied in an experiment or observational study. At issue is whether the treatment has an effect on the outcome or variable of interest.

■ **treatment effect**

the effect of the treatment on the variable of interest. Establishing whether the treatment has an effect is the point of an experiment.

■ **treatment group**

the individuals who receive the treatment, as opposed to those in the control group, who do not.

■ tree

a tree is a graph with the property that there is a unique path from any vertex to any other vertex traveling along the edges.

■ triangle

a three-sided figure that can take several shapes. The three inside angles add up to 180o. Triangles are divided into three basic types obtuse, right and acute; they are also named by the characteristics of their sides equilateral, isosceles, and scalene. The area of a triangle is 1/2 x perpendicular height x base.

■ triangular number

a number of the form n(n+1)/2.

■ triangulation

a tiling of some object such as a manifold by simplices.

■ trigonometric ratios

the ratios of the lengths of pairs of sides in a right triangle, i.e., sine, cosine and tangent.

■ trigonometric ratios

the ratios of the lengths of pairs of sides in a right triangle, i.e., sine, cosine and tangent.

■ trigonometry

the branch of mathematics involving triangles that combines arithmetic, algebra and geometry. Trigonometry is used in surveying, navigation and physics.

■ trinomial

an algebraic expression consisting of 3 terms.

■ triple point

any kind of point where three things meet: Three surfaces in space, three phases of a substance in a phase diagram, three shock waves, etc.

■ trivalent graph

a graph such that each vertex has three edges.

■ tromino

a three-square polyomino.

■ truncated pyramid

a section of a pyramid between its base and a plane parallel to the base.

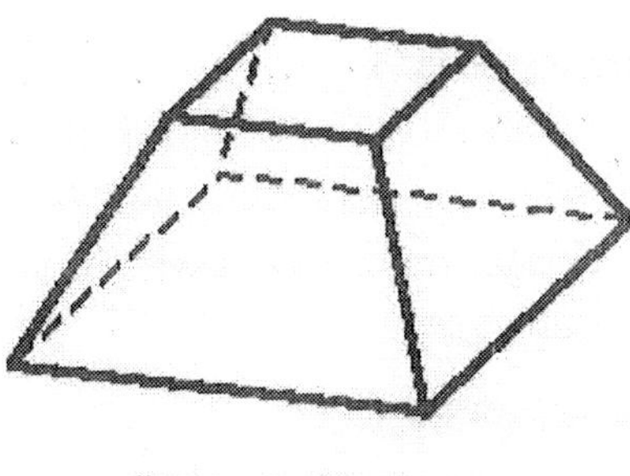

■ **tuple**
an ordered list of elements from a set, usually represented as (a1, a2, a3, ... ,an). (Sometimes angle brackets are used in place of parentheses.) A tuple with only two elements is called an ordered pair, and tuples with 3, 4, and 5 elements are called ordered triples, quadruples, and quintuples, respectively. In general, a tuple with n elements is called an n-tuple. A relation on a family of sets is represented by a set of tuples.

■ **turing machine**
a mathematical model of a computer consisting of an automaton travelling along a tape. The automaton at any given time is in some state depending on its previous state and the data at its current position along the tape, and its state also determines whether it moves down the tape and what it writes to the tape at its current position.

■ **turn symmetry**
when an object matches itself when rotated about a point.

■ **twin primes**
two prime numbers that differ by 2. For example, 11 and 13 are twin primes.

■ **two-dimensional figure**
has only length and width, like a square drawn on paper.

■ **type i and type ii errors**
these refer to hypothesis testing. A Type I error occurs when the null hypothesis is rejected erroneously when it is in fact true. A Type II error occurs if the null hypothesis is not rejected when it is in fact false. See also significance level and power.

■ **unbiased**
not biased; having zero bias.

uncontrolled experiment
an experiment in which there is no control group; i.e., in which the method of comparison is not used the experimenter decides who gets the treatment, but the outcome of the treated group is not compared with the outcome of a control group that does not receive treatment.

uncorrelated
a set of bivariate data is uncorrelated if its correlation coefficient is zero. Two random variables are uncorrelated if the expected value of their product equals the product of their expected values. If two random variables are independent, they are uncorrelated. (The converse is not true in general.).

uncountable
a set is uncountable if it is not countable.

unilateral surface
a surface with only one side, such as a Moebius strip.

unimodal
a function is unimodal if it goes up and then it goes down, once. For example, a bell curve.

unimodular
a square matrix is unimodular if its determinant is 1.

union
the union of two or more sets is the set of objects contained by at least one of the sets. The union of the events A and B is denoted "A+B", "A or B", and "AUB". C.f. intersection.

unit
1.a standard measurement.
2. a member of a population

unit circle
a unit circle is a circle with radius 1.

unit cube
a cube with edge length 1.

unit fraction
a fraction whose numerator is 1 (e.g., $1/\pi$, $1/3$, $1/x$). Every nonzero number may be written as a unit fraction since, for n not equal to 0, $n = 1/(1/n)$.

unit rate
a rate with a denominator of 1.

unit square
a unit square is a square of side length 1.

unit vector
a vector of length 1. It is customary to designate i as the unit vector in the x direction, j as the unit vector in the y direction, and k is the unit vector in the z direction.

unitary divisor
a divisor d of c is called unitary if gcd(d,c/d) = 1.

unity
one.

univariate
having or having to do with a single variable. Some univariate techniques and statistics include the histogram, IQR, mean, median, percentiles, quantiles, and SD. C.f. bivariate.

validity
an argument that is correctly inferred or deduced from a premise.

valuation
in convex geometry, a function f on convex sets such that f(A) + f(B) = f(A cap B) + f(A cup B).

variability
numbers which describe how spread out a set of data is (for example, range and quartile).

variable
a numerical value or a characteristic that can differ from individual to individual. See also categorical variable, qualitative variable, quantitative variable, discrete variable, continuous variable, and random variable.

variance, population variance
the variance of a list is the square of the standard deviation of the list, that is, the average of the squares of the deviations of the numbers in the list from their mean. The variance of a random variable X, Var(X), is the expected value of the squared difference between the variable and its expected value

$Var(X) = E((X - E(X))2)$. The variance of a random variable is the square of the standard error (SE) of the variable.

variation

in analysis, the amount that a function increases plus the amount that it decreases.

vector

quantity that has magnitude (length) and direction. It may be represented as a directed line segment.

vector product

the vector product (also called cross product) of two vectors u and v, denoted u × v and called "u cross v," is a vector w whose magnitude (length) is the product of the magnitudes of u and v and the sine of the angle between them, and which points in a direction perpendicular to the plane containing u and v so as to form a right-handed system, as in the figure.

note that the directedness of the vector product implies that it is not commutative.

vector space

a structure consisting of two kinds of elements called scalars and vectors, with operations of addition of pairs of scalars or pairs of vectors, and multiplication of pairs of scalars or a scalar and a vector. The vectors form an Abelian group under addition, and the scalars form a field under their operations, and the vector space is said to be over that field.

if the scalar field is the real numbers or the complex numbers and the vectors are in n-dimensional real or complex space, then the space is called an n-dimensional real or complex vector space accordingly. Multiplication of vectors by scalars is associative with scalar multiplication, and distributive over both scalar addition and vector addition. Symbolically, for scalars a, b, and vectors u, v, vector spaces are usually denoted by V, and it is conventional to write the scalar on the left of a scalar multiplication. When there is any possibility of confusion, the vectors of a vector space are usually specially marked, either by drawing a (right pointing) arrow over them or by writing them in bold face.

■ **velocity**

the rate of change of position. The first derivative of the position function.

■ **Venn diagram**

a pictorial way of showing the relations among sets or events. The universal set or outcome space is usually drawn as a rectangle; sets are regions within the rectangle. The overlap of the regions corresponds to the intersection of the sets. If the regions do not overlap, the sets are disjoint. The part of the rectangle included in one or more of the regions corresponds to the union of the sets. This page contains a tool that illustrates Venn diagrams; the tool represents the probability of an event by the area of the event.

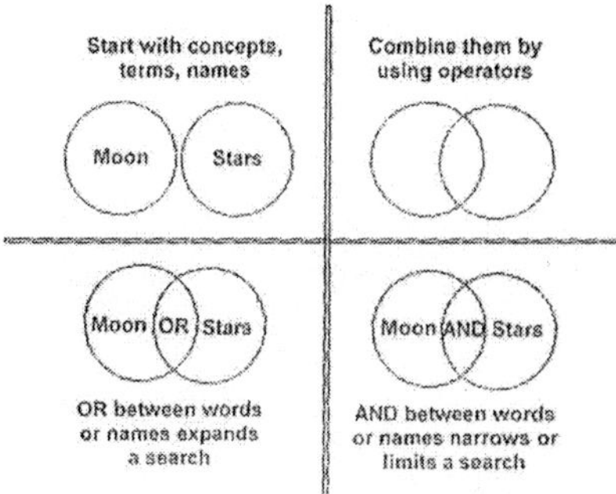

■ **Verma module**

a linear representation of a Lie group generated by applying ladder operators to a certain starting vector, called a highest-weight vector, in the same way that one constructs spin states of a particle in quantum mechanics starting from a state of maximal spin in the z direction. Unlike the set of spin states, a Verma module has no lowest-weight vector and is necessarily an infinite-dimensional vector space.

■ **vertex**

the corner of a geometrical figure, can be formed by lines, planes, etc.

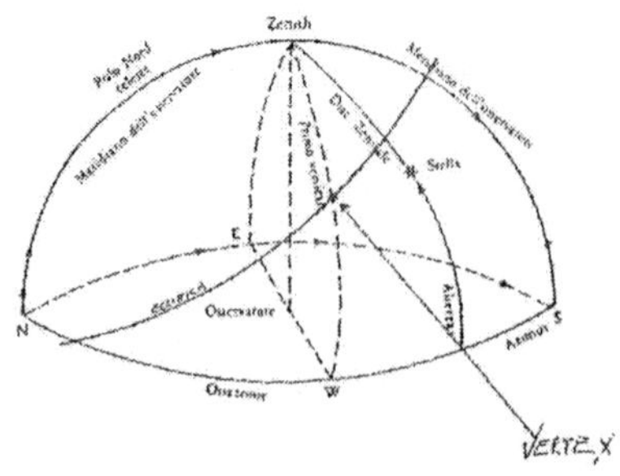

■ **vertex operator algebra**

an algebra with an associative and commutative addition operation and a non-associative, parameter-dependent multiplication operation. The operations satisfy axioms related to conformal structures on surfaces. Vertex operator algebras are a mathematical abstraction of 2-dimensional quantum field theories.

■ **vertical angle**

two angles with equal measures formed by intersecting lines.

■ **vigesimal**

related to intervals of 20.

■ **vinculum**

the horizontal bar in a fraction separating the numerator from the denominator.

■ **Virasoro algebra**

a modified version (technically a central extension) of the Lie algebra of tangent vector fields (or infinitesimal motions) of a circle. It was first described abstractly in the physics literature by commutation relations in much the same way as rotation operators are discussed in quantum mechanics. It appears in the study of vertex operator algebras.

■ **volume**

the measure of space occupied by a solid body.

■ **vulgar fraction**

a common fraction.

■ **weak inequality**

an inequality that permits the equality case. For example, a is less than or equal to b.

■ **weak shock wave**

a shock wave with a relatively small discontinuous jump in pressure or another parameter.

■ **weak/weakly**

in mathematics and especially in analysis, an object is called weak if it is of a generalised kind with fewer properties, and a property holds weakly if it holds in a lesser sense. For instance, a weak solution to an equation might be a discontinuous solution if a straightforward interpretation implies continuity.

Weierstrass approximation theorem

a foundational theorem that, given a smooth function, one can find a polynomial whose values and derivatives are arbitrarily close to those of the function.

wff

a well-formed formula.

whole number

any of the following numbers.

{0, 1, 2, 3, 4, 5, ...}.

winding number

the number of times a closed curve in the plane passes around a given point in the counterclockwise direction.

witch of Agnesi

a curve whose equation is x2y=4a2(2a-y).

X

roman numeral for 10.

x-axis

the horizontal axis in the plane.

x-intercept

the point at which a line crosses the x-axis.

xOR, exclusive disjunction. xOR is an operation on two logical propositions If p and q are two propositions, (p XOR q) is a proposition that is true if either p is true or if q is true, but not both (p XOR q) is logically equivalent to ((p OR q) AND NOT (p AND q)).

x-pentomino

a pentomino in the shape of the letter X.

Yang-Baxter equation

an algebraic relation arising in statistical mechanics, topological quantum field theory, and quantum groups in which two tensors, one naturally represented by a right-side-up triangle and the other by an upside-down triangle, are equal.

■ **yard**

a measure of length equal to 3 feet.

■ **y-axis**

the vertical axis in the plane.

■ **year**

a measure of time equal to the period of one revolution of the earth about the sun. Approximately equal to 365 days.

■ **y-intercept**

the point at which a line crosses the y-axis.

■ **z statistic**

a Z statistic is a test statistic whose distribution under the null hypothesis has expected value zero and can be approximated well by the normal curve. Usually, Z statistics are constructed by standardizing some other statistic. The Z statistic is related to the original statistic by.

z = (original - expected value of original)/SE(original).

■ **z transform**

given a discrete signal x(n), the z-transform of x is.

z { x (n) } = X(z) = ?n=- 8 8 x(n)z-n

■ **Zeno of Elea**

greek philosopher famous for certain paradoxes, including the Paradox of the Arrow, the Paradox of the Tortoise and Achilles, and the Paradox of the Moving Rows. These constitute the earliest known arguments from infinity.

■ **Zeno's Paradox of the Arrow**

one of the famous paradoxes of Zeno of Elea. Consider an arrow in flight. At each moment of time the arrow may be considered to occupy a specific region in space. This poses the following problem: if the arrow is in a particular point of space in any given moment, then how is it moving? And if it is not moving, how does it get from point to point in the passage of time?

■ **zepto**

a prefix meaning 10^{-21}.

■ **Zermelo Fraenkel set theory**

the standard set theory in which most mathematics is formalized. Its axioms include the pairing axiom, the power set axiom, the axiom of infinity, the axiom of extensionality, the axiom of replacement, the separation axiom, the union axiom, and the foundation axiom. Abbreviated ZF. When the axiom of choice is assumed, this theory is abbreviated ZFC.

■ **zero**

0

■ **zero divisors**

nonzero elements of a ring whose product is 0.

■ **zero element**

the element 0 is a zero element of a group if a+0=a and 0+a=a for all elements a.

■ **zeros of a function**

the values of the domain of a function which are mapped on to zero or, less formally, the points at which the value of a function is zero.

■ **z-intercept**

the point at which a line crosses the z-axis.

■ **Zipf's law**

if frequency F is plotted against rank i, the relationship between the two is of the form F~1 / i a , with a close to 1; ie there is a power-law relationship. This law holds for many naturally-occurring populations; for example the frequency distribution of words in texts.

■ **zone**

the portion of a sphere between two parallel planes.

■ **zoned decimal**

a method, in computing, of representing decimal numbers by using one byte to represent each decimal digit. The last byte represents both the last digit and the number's sign.

■ **z-score**

the observed value of the Z statistic.

■ **z-test**

an hypothesis test based on approximating the probability histogram of the Z statistic under the null hypothesis by the normal curve.

NOTES

NOTES

NOTES

Other Books on

LOTUS ILLUSTRATED DICTIONARIES

1. First English Dictionary **(New)**
2. Astronomy
3. Agriculture
4. Anthropology
5. Archaeology
6. Architecture
7. Art
8. Banking Finance & Accounting
9. Bio - Chemistry
10. Business Administration
11. Bio - Technology
12. Biology
13. Botany
14. Culture
15. Chemical Engg.
16. Chemistry
17. Civil Engineering
18. Computer Science
19. Commerce
20. Cooking & Food
21. Ecology
22. Economics
23. Education
24. Electrical Engineering
25. Electronic & Telecommunication
26. Environmental Studies
27. Festival
28. Geography
29. Geology
30. Genetic Engineering

31. Health & Nutrition
32. History
33. Import and Export
34. Information System Management
35. Internet
36. IT
37. Inorganic Chemistry
38. Law
39. Library & Information Science
40. Literature
41. Management
42. Mathematics
43. Marketing & Sales
44. Mass Communication
45. Mechanical Engg.
46. Medical
47. Music
48. Organic Chemistry
49. Philosophy
50. Physical Education
51. Physics
52. Psychology
53. Science
54. Sex
55. Sociology
56. Sports
57. Textile
58. Zoology
59. Dictionary of Veterinary Sciences **(New)**
60. Dictionary of Synonyms and Antonyms **(New)**
61. Dictionary of Idioms and Phrases **(New)**

Unit No. 220, 2nd Floor, 4735/22,
Prakash Deep Building,Ansari Road, Darya Ganj,
New Delhi- 110002